W9-AXQ-414

INTELLIGENT INNOVATION

FOUR STEPS TO ACHIEVING A COMPETITIVE EDGE

JOHN A. COGLIANDRO

ISBN-10: 1-932159-61-4
ISBN-13: 978-1-932159-61-5

Printed and bound in the U.S.A. Printed on acid-free paper
10 9 8 7 6 5 4 3 2 1

Library of Congress Cataloging-in-Publication Data

Cogliandro, John A.,
Intelligent innovation : four steps to achieving a competitive edge / by John A. Cogliandro.
p. cm.
Includes index.
ISBN-10: 1-932159-61-4 (hardcover : alk. paper)
ISBN-13: 978-1-932159-61-5 (hardcover : alk. paper)
1. Technological innovations. 2. Creative ability in business. 3. Organizational change. 4. Success in business—Case studies. I. Title.
HD45.C534 2006
658.4′063—dc22 2006025677

Direct all inquiries to J. Ross Publishing, Inc., 5765 N. Andrews Way, Fort Lauderdale, FL 33309.

Phone: (954) 727-9333
Fax: (561) 892-0700
Web: www.jrosspub.com

Dedicated to my parents, who provided the inspiration to make this happen; to my father's dedication to excellence in everything God, family, and country that set the example for my life; and to my mother's wit, beauty, and caring that kept us all going during good times and bad.

To my wife, who edited and contributed to this book through countless conversations, white board sketches, what-if scenarios, and after-dinner ramblings.

And to Don, who gave me a chance in Piqua.

TABLE OF CONTENTS

PREFACE

Intelligent Innovation™ is not a method. It is not a get-rich-quick scheme or something that happens instantly after a consultant conducts a break-out session. It is not something that happens at a point in time. Instead, Intelligent Innovation is a life cycle multidimensional process backed up by a culture of strategically balanced, creative problem solving, an intuitive sense of the market, and a drive to succeed. It is a can-do attitude in everything people, process, and productivity. In this book, it is described as having four distinct phases distilled from researching hundreds of success stories in multiple industries in several countries. The process phases of a turbine engine—intake, compress, combust, and thrust—fit our life cycle innovation model perfectly. Intelligent Innovation and subsequent success are a result of doing many tasks better, including management, motivation, finance, asset allocation, decision making, and authority-responsibility coordination. Ideally practiced seamlessly and iteratively, Intelligent Innovation is applied in a phased approach, with the concepts in this book ultimately adding thrust and creative energy to any organization.

Whereas the current popular literature on the subject of innovation concentrates on the front end of the business, accenting the creative process, Intelligent Innovation is much more than that; it requires a cradle-to-grave process in order for an organization to capitalize on early-phase ideas. In short, Intelligent Innovation seeks to prove, encourage, explain, and provide instruction relating to three basic principles:

1. Innovation must be pervasive throughout the organization for blockbuster success. It must be discovered, encouraged, used, and capitalized on in every phase of development from the early concept

phase through delivery and support. In short, a life cycle approach to innovation is required.

2. Innovation is necessary for corporate value (stock) growth, regardless of all other influencing factors. Corporate growth cannot occur without consistent innovation in product development.
3. Innovation transcends traditional boundaries. It must be in product, process, policy, and people. Innovation in one area but not another is suboptimal.

Thomas Alva Edison's definition of genius as "1 percent inspiration and 99 percent perspiration" has became so well known it is now cliché. Having been a moderately successful inventor myself (read: I made a few cool inventions but never made a killing on them), I understood in particular the inspiration element, but I knew there was more to success than a good idea and hard work. Where does success come in? Was I missing something in the overall product development process—some key to wild blockbuster success? I have also been involved in countless development projects, helping bring to market hundreds of products, from civilian and military aircraft to shock-absorbing snow shovels and folding bicycles. Some of these products were successful and delivered on their promises of financial return and consumer performance. Others were not. Some delivered returns immediately to the consumer, while for others, returns came to the developer and investors only after years of pain and debt. I wondered, *was there some magical formula the blockbusters had that the mediocre performers did not?*

I set out to answer this by reviewing a series of formal and informal research studies and observations, which are included in this book. While looking at the data for a magic formula, a missing key, a pattern began to emerge. The pattern showed four clear phases that the blockbusters had followed—phases during which the innovative process was used to pass on to the next phase. These four phases were interconnected and managed as a whole. Intertwined with these phases were the three principles of Intelligent Innovation—along with some rather counterintuitive keys.

In this book I describe the innovative factors, tools, and methods that the blockbuster producers used to achieve success, as well as the phases of development—the life cycle view of innovation that any organization must use to achieve that success. Innovation in one form or another in every phase was the link among the blockbusters and what distinguished them from all the rest.

Innovation in these studies is not restricted to the fluffy, up-front "ideation" that many consulting firms purvey. Ideation will get you precisely to the end of the session and no further. Innovation comprises patents and intellectual property, but it is not limited to those areas. It is the insight, the problem solving, and the realization of unique opportunities throughout the entire product life cycle (including the fun, creative front end) that make the blockbusters what they are. One of the best examples of this early ideation versus consistent innovation conflict is often found in Donald Trump's TV show, "The Apprentice." Often, the team that wins that week's test is not the team with the coolest, most innovative idea; it is the team that puts it all together: wow marketing, clever distribution, sales, innovative product, innovative process, location, and so on. While Mr. Trump often celebrates the most innovative idea, the winning team is always the one that made the most money, regardless of the method or actual idea. In effect, Trump's final measure is the overall measure of the life cycle process output.

In this book, each chapter is designed to provide specific topical value and to fit into an overall process improvement paradigm using the four process phases of success: intake, compress, combust, and thrust. The reader may want to review Figures 2-6 and 2-7 in Chapter 2 prior to reading the rest of the book, as they overlay these phases with a typical product development process. For a quick read, I recommend reading Chapters 1, 2, 4, 6, 10, 15, and 16 and the Afterword.

While much of this book concerns product development, process and service industries will also benefit, as there is ample proof that the principles of blockbuster success and innovation throughout the entire life cycle apply to these industries as well. Whatever your industry, I hope that Intelligent Innovation becomes a pervasive attitude and culture, applied in every facet and by everyone in your organization, rather than a process or single-point solution.

ABOUT THE AUTHOR

John Cogliandro is a program manager with the Mission Innovation Cross Business Team (MI CBT) of Raytheon Integrated Defense Systems and president of Intelligent Innovation Management Consulting. For Raytheon, he was appointed to the MI CBT in September 2004 after serving in various leadership roles for the development of the Next Generation Surface Combatant, or DD(X), program for the U.S. Navy. Currently he contributes to the development of capabilities in intellectual property licensing, System of Systems (SoS) performance optimization, and mission assurance.

Prior to Raytheon he served as the program manager for the F-18 Pod and Airborne Laser guidance systems and was the systems engineering lead for the CVN-X aircraft carrier program.

Through his company Intelligent Innovation, Cogliandro is an internationally recognized speaker and author, having published over 20 articles on technology development, risk management, intellectual property strategy, and strategic planning. He is the original developer of 3D Risk Management (patent pending) and The I-Quotient Self-Evaluator. He specializes in developing innovation strategies for organizations of all types and speaks regularly around the world on the topic. His favorite keynote address is titled "Solving the Risk vs. Innovation Dilemma"—an upbeat, challenging talk filled with popular anecdotal stories, easy-to-apply concepts, and humor.

Cogliandro holds numerous certifications, degrees, awards, and chair positions. He has earned a Bachelor of Science in Industrial Engineering and Master of Business Administration. Cogliandro also has certificates in Intermediate and

Advanced Systems Planning and has also been certified in Systems Integration by the Institute of Industrial Engineers.

ACKNOWLEDGMENTS

With special thanks to:

Drew, who put up with me in the early days of disjointed thought and listened with an amazing ability not to become sick while providing intelligent responses.

Chris, whose advice changed my life for the better.

Craig, whose sharp wit was an odd source of motivation.

David, whose faith in God and encouragement go hand in hand, and who makes my life and the world a better place.

Alan, the smartest guy I know, who leads by example.

Johnny, the sales guy, who encourages people with everything he says.

My sister, whose example of warmth and giving is unmatched in this world.

The Babo and Cincy gangs, Dave, Kami, Gladys, RJ, Keith, Sue, John, Paul, Linda, Kathleen, and so on.

Brian and Matt, who taught me systems engineering.

Francine, Linda, Lawrence, Dave, Matt, and all the gang.

Steve, a talented engineer who helped get this book published and who taught me risk management.

Quinn and Kristen, who walk by character first, accomplishments second.

Dan and Lee, who practice what others only dare to theorize.

And Jonathan, who provided editing and insight.

Psalm 127:1 *Unless the Lord builds the house, the builders labor in vain.*

LIST OF FIGURES

At J. Ross Publishing we are committed to providing today's professional with practical, hands-on tools that enhance the learning experience and give readers an opportunity to apply what they have learned. That is why we offer free ancillary materials available for download on this book and all participating Web Added Value™ publications. These online resources may include interactive versions of material that appears in the book or supplemental templates, worksheets, models, plans, case studies, proposals, spreadsheets, and assessment tools, among other things. Whenever you see the WAV™ symbol in any of our publications, it means bonus materials accompany the book and are available from the Web Added Value™ Download Resource Center at www.jrosspub.com.

Downloads available for *Intelligent Innovation: Four Steps to Achieving a Competitive Edg*e consist of a FEOTB requirements checklist, methods for defining corporate core competencies, a business development process model and rating chart, and an innovation self-evaluation tool.

1

INNOVATION: THE KEY TO SUCCESS

> "Great moments are born from great opportunity."
> —***Coach Herb Brooks***, *1980 U.S. Olympic Hockey Team*

We all have an intuitive sense that innovation is important. We know it must be a contributor to the success we collectively trumpet, but how much? Where does it come into the picture? Is innovation just for product companies? Is it just for great musical composers? Architects? How about the trucking firm working on improving on-time delivery while lowering fleet fuel costs? How about a regional airline working on an initial public offering (IPO)? How about a computer manufacturer working on a long-life laptop battery? Does it relate to new products, things we touch, or does it relate to processes that get new products or services to the market faster and at a higher level of initial quality?

Do any of these situations "need" innovation? Do they all? If so, at what point do they "innovate"? Is the innovation a predetermined act, a subject of the will, or is it something much more complex, such as a subconscious eruption of successful thought?

In this text, we ask and answer these questions: What is innovation? Is it an action (verb, i.e., to innovate) or a thing (noun)? Who is it for? When do you use it, and how?

"Wow!" I exclaimed to the sleeping woman next to me on the plane. She did not respond to the tired-looking man with the outdated laptop. There it was, right in front of me on that little screen, indisputable proof that

systems engineering and its close relative industrial engineering, along with their basic decision-making paradigms, were having a positive, measurable impact on the companies that followed their doctrines. It was like finding the ark, or seeing the Red Sox win the World Series . . . a revelation of epic proportion. OK, calm down, I told myself. Finally, she looked over at me and said, "Engineering is for geeks. Innovation is more important." I said, "They are intertwined, synonymous, codependent." "Prove it," she said while rolling over to sleep again . . .

The proof of a link between decision management and performance and several other core innovation and process principles for this book came from a 2002 landmark study performed by Strategic Balancing Management Consulting of Bedford, Massachusetts. The results of that study have been verified and validated by the results of several other studies that reached similar or supportive conclusions. Strategic Balancing Management Consulting's study took several years, spanned eight countries and three continents, and included over 100 interviews and surveys from some of the world's most revered firms from the European Community, Australia, United States, Canada, and Mexico; they included Mercedes-Benz, Ford, Boeing, NASA, General Electric, the Nuclear Regulatory Commission, Analog Devices, MIT, and many others. The study also included countless smaller nondescript firms—that "silent infrastructure" that the business schools and consultants long ago forgot.

The study was commissioned after a particularly intense risk management training session performed for Technology Training & Consulting (TTC), Incorporated, where an audience member questioned the validity of risk management and then the validity of any management. In short, he asked for proof of results. His questions were designed to be provocative and controversial, not adversarial, but they were largely unanswerable with the current data available in the marketplace. A series of 15 questions were subsequently developed by Strategic Balancing Management Consulting's research division (excerpt available in Appendix A). The study became the Strategic Risk Management (SRM) Survey. The questions were aimed at measuring aspects of management and strategy. It focused on three broad categories: (1) decision-making efficiency (linked to requirements basis), (2) determining an innovative quotient (termed I-Quotient™, an objective grade as to a company's ability to promote and capitalize on innovation to determine the potential for innovative yield), and (3) understanding the use of vision and mission in management (a hidden measure of management focus, employee stress, and organizational health) of the respondents. Furthermore, the data in the SRM Survey allowed firms to be categorized into a demographic matrix construct, such as large versus small, manufacturing versus service, and so forth.

The results were astounding, and some were quite counterintuitive. Some factors that led to success were quite logical, such as "clear decisions and employee empowerment were good for overall performance." But other factors in the data were surprising and flew in the face of common sense and accepted project management techniques. And yet these odd findings stood the ultimate test of validity: marketplace profits. In an unrelated study with supporting conclusions, The Boston Consulting Group's 2005 Innovation Survey of Top Executives stated that "90 percent of the executives said that generating organic growth through innovation is essential for success."

Remember those questions that were asked in the first paragraph? Where do you fit in? Are you an architect, nurse, inventor, manager, attorney, chief technical officer, senior engineer, technical director, or software programmer? The initial SRM Survey focused on firms that perform very visible product development activities. They design and make cars, planes, and spaceships. The primary study was later expanded to include over 1,000 observations of less tactile organizations, including service industry personnel (nurses, lawyers, and architects) and service products such as insurance, coffee shops, and so forth. These results were included in the Extended SRM (ESRM) Survey.

Both sets of research point to the same results:

- Innovation is important in all aspects of business.
- Innovation refers to both an action (to innovate) and a thing (an innovation).
- Innovation is both a process and a point of inspiration.

More specifically, the SRM study produced data suggesting several key points, which are outlined here and discussed in detail throughout the book. They are broadly categorized into five areas.

MANAGEMENT AND DECISION-MAKING ABILITY

The study showed that some 76 percent of the respondents viewed a lack of decision-making clarity and efficiency—suggested by multiple, unlinked approval cycles and reworked decisions—as a key aspect of deterring performance improvement. The comments suggested that organizational policies and procedures that supported the practices of good decision making were badly needed. In contrast, some 24 percent of the firms reported adequate or efficient decision making. And not surprisingly, these same firms often reported high performance and clear dissemination of a corporate vision and mission. Ultimately the study found that these same measures were related to the ability to capitalize on innovations (both internal and external), and the I-Quotient became a reliable measure of the ability to achieve yield on any innovation. This area of data became our engine control module, explained later in the turbine engine analogy used throughout this book.

DISSEMINATION OF VISION AND MISSION

Clarity and thorough dissemination of a single organizational vision and mission were key aspects of success in 24 percent of the respondents. Much more telling was an observation the data input clerk noted while tabulating the survey forms. She noticed that in no instance did a respondent ever answer yes to both of the following two questions:

- Does your organization have a clear vision and mission?
- Do people understand how the vision and mission relate to their jobs, including new product development?

Even the high-performing 24 percent minority who had a clear vision in their firms did not check the box indicating that the vision or mission was adequately disseminated, understood, and used in subsequent management decisions or operational activities. These firms were close but were lacking a common ideological drive to succeed amongst the entire team. They had failed to translate and communicate the visions and goals into the vernacular of every employee. This translation helps transform employees from workers to owners.

CORRELATION BETWEEN VISION AND THE RESULTING ACTIONS

The vision–action mismatch mentioned previously made me suspicious. I wondered, if these firms were the "good" ones, what made the great ones? What made some projects, products, and even entire organizations stand out as global success stories? Looking at the data more closely, I could only find more clues as to how not to do it, clues to things that were broken. Many respondents had indicated that complicated requirements and justifications were needed to obtain research and development (R&D) funds or, worse, to keep R&D funds after they had been allocated. Many others indicated in the written notes that their organizations had major "alignment" problems and that performance was suffering. Of interest was the word "alignment"; it was used over and over again in the survey by respondents from different firms in different industries. In fact, it became the basis for Chapter 8 of this book, "The Strategic Balancing Method". And yet some of these organizations were responsible for at least one major market success at one point in their history.

My subsequent research and observations led me to believe this phenomenon was masking a more powerful concept, which I dubbed *ideological ingenuity*. Years ago I read a historical account of the taking of the hill on Iwo Jima during the end of World War II. Both sides, over 23,000 U.S. Marines and over 21,000 Japanese soldiers, fought with ferocity and died with honor. The fighting was fierce, and from the Allied point of view, it was necessary to repel tyranny and provide a base of operations for U.S. bombers. Yet more than any bullet or bomb, the event that turned the tide was the U.S. commanders' goal to put a flag on the hill. The hill, Mount Suribachi, was defended by Japanese who had dug in and had every tactical advantage. Once Mount Suribachi flew the American flag, it became an icon for the entire war, and the famous image of the men raising the flag there helped the United States dominate world economies and politics for the next 40 years.

What was it about that flag? What was it about that singular icon and the goal it represented that had such power? Aspects of the story sounded so familiar. It sounded just like the story of the development of the Ford Taurus and the Lockheed SR-71. It sounded like the development of the Mazda Miata and the Ferris wheel at the 1902 World's Fair. The key here was not the flag, but the focus, energy, determination, and ingenuity that a single ideological goal provides to a team of employees or soldiers. In all these cases, a small team of people, pitted against huge obstacles with impossible goals, united and pounced on the common goal. A famous quote from Admiral Chester W. Nimitz explains this well: "Among the Americans who served on Iwo Island, uncommon valor was a common virtue."

The ESRM study revealed that projects known for saving the company, the country, or lives and for being wildly successful shared a team ideology more frequently than any other characteristic. Some of the characteristics they did not have in common included individual smarts, technological breakthroughs, management techniques, finances, or corporate structure. These were the results that ran "counter to accepted practice" from the results of the study. There was absolutely no correlation between success and flat, nimble management structures or hierarchical, traditional management structures. It was as though all the pontificating of business schools, consultants, and talking heads had no meaning compared to the power of a common ideological goal. Likewise, adequate or inadequate finances had little correlation to success—definitely a counterintuitive result.

The ESRM study looked at many characteristics of a project's success. Two questions were asked of the participants that were aimed at uncovering the driving forces behind a project. One question asked if projects were usually constrained or resources limited in some way (schedule, budget, manpower), what was the impact on their relative success? The consistent answer, backed up by hundreds of examples, was that resource constraints (and the resulting pressure on the development team) often helped, not hindered, the project. These concepts form the basis for several chapters on resource management, planning, and process innovation.

UNDERSTANDING THE NEED FOR INNOVATION AT DIFFERENT PHASES OF A PROJECT

Understanding the phases of and need for innovation and risk in every aspect of product development and the organization's activities was lacking in well over 80 percent of the firms that reported. In only two instances did respondents indicate that innovation was understood or expected in the later stages of development. That is like saying a master chef can go to a market, buy all the finest ingredients, and create a wonderful new recipe through trial and error, and then on the most important day of his career, the opening of a new restaurant, he throws it all together and serves it raw on a paper plate. The final stages of any product or service construction and delivery are equally important or perhaps more important to the customer. The notion of innovation throughout the phases, which I dubbed the *Innovation Lifecycle,*™ forms the basis for several chapters. The life cycle concept is supported by anecdotal stories from subsequent interviews with participants in the ESRM study. The relationship between innovation needs and phases is discussed in Chapter 14, "Solving the Risk vs. Innovation Dilemma."

CONSISTENT FUNDING OF RESEARCH AND DEVELOPMENT

The SRM study included several questions about the funding and management of R&D. The hypothesis here is that most innovation and product development has some start or link to an organization's research department. The SRM study aimed to uncover a link between the management of R&D and the outcome. The R&D-oriented data formed a sort of bell curve, skewed to the left. The "left" would mean that R&D projects or monies sometimes required justification for continued funding and must reapply yearly. This is in sharp contrast to having R&D funding linked to milestones, which would be a skew to the right. A number of respondents indicated no need to reapply for funding, and not surprisingly, these firms typically also noted that they were more tolerant of risk and had a broader view of innovation and the need for innovation. When an organization funded (with money and other resources) innovation from its birth point—wherever that was—to its delivery to the marketplace, a strong link with success was generally, but not always, noted. This trend led to the concepts explored in Chapter 15, "Stocking Up on Innovation." The chapter examines the link between stock price, which is the ultimate measure of success, and innovation.

Funding Mismatch

There was a mismatch between the "point" at which innovation was expected (or needed) within the organization and the funding of those activities. A whopping 90 percent of the respondents claimed that their firm's predominant expectation of innovation was in the early (R&D) phase or in the engineering phase of a product's life cycle; however, management or policy made the funding at that point difficult to attract, spend, and keep. Furthermore, those respondents noted that the middle phases of a product's development, design, production, and delivery were largely ignored when it came to innovation expectations. This is one of the most important factors preventing many firms from achieving much higher levels of success in product delivery, financial performance, and ultimately, stock price.

I'll explain these points and the details of the two studies in further detail in subsequent chapters. For now, refer to the sidebar on the next page that discusses Jane's Toy Company, an actual case study from my research. This simple case study demonstrates the continuous nature of and need for innovation.

Much has been written about the importance of innovation. This book in particular attempts to clarify and aid the understanding and application of innovation, using practical examples and relevant cases. Each chapter introduces a topic and provides both theoretical and practical examples and concludes with an application guide. The examples are all taken from real situations, mostly from current-day client interactions, news stories, the SRM studies, and my own primary research.

The book covers five main topics:

1. Innovation and the Innovation Lifecycle
    - The importance of innovation
    - Understanding the Innovation Lifecycle for your particular organization and industry
    - The four phases of innovation and new product development
    - Sparking innovation: understanding the difference between chickens and eggs
2. Managing innovation for increased yield

CASE STUDY: JANE'S TOY COMPANY

Jane was working on a new children's toy. She had developed it after the violent tragedy of 9/11 in New York City. That day she saw several children in the airport who were traumatized by the terrorist attack. Her mind put this need together with her experience as a caregiver to children in a hospital. She frequently used different toys to care for the children who were suffering from a variety of ailments. She thought to herself, *What they need is a toy that soothes, a toy that helps them communicate their feelings and anxiety. Then they will feel better.* This was her "eureka" moment. The toy was innovative, combining several design elements, textures, and features that allowed a child to communicate feelings to a guardian through the toy. Six months later Jane was still inventing. There were materials to develop, creative distribution agreements to forge, funding streams to uncover, and so on. She was surprised to find that the invention process has a beginning but no end. It continues through every aspect of a business, from design to procurement to packaging to distribution.

Jane now owns a successful children's specialty toy business, four years after her "innovative invention" moment. In Jane's case, innovation was critical at each step in the business process. Her lesson was that the actual toy's design and features were the easiest and least "innovative" part of the process in the tough, hypercompetitive toy industry.

- The integration of good decision making as a key to implementing innovation
- The integration of good resource allocation as a key to implementing innovation
- Process innovation versus product innovation
- Policy: the ultimate enhancer or killer of innovation

3. Valuing innovation
    - Innovation is a necessary value enhancer, driving corporate valuation and public perception, capitalizing on innovation
    - The natural tension between risk and return, between linear and exponential innovation, can and should be solved at a strategic and operational level prior to proceeding with any development project
4. Innovative strategies and business models
    - Innovative business constructs and business models can be (and often are) more powerful and profitable than any single new product innovation

    - Strategic planning is the ultimate catalyst in the Innovation Lifecycle, amplifying the effects of innovation at every step when done correctly
5. Adopting Intelligent Innovation
    - Understanding that innovation is an attitude (both an action and thing, a verb and a noun) and should be adopted throughout the entire organization
    - How to adopt the innovation attitude, considering how cultural issues are always harder than process issues, and some practical ways to improve them

Note that this entire book is written to be applied to almost any organizational situation, both product and service. While many of the examples are for hard products, the service industry's need for innovation is just as valid, and perhaps even more sensitive to its application and implications. For example, a business construct, a "soft" non-product-oriented idea, came in the form of a city reconstruction effort called Coney Island. Coney Island, New York, was developed in 1923 at the cost of $4 million, a huge sum for the cash-strapped city. While numerous physical innovations were required to dredge the harbor and build the boardwalk, the business model was truly a breakthrough, generating taxable returns that paid for the project in one year! The social and socioeconomic returns to the local population, which lasted for generations, were immeasurable.

Before we continue, some common definitions may be helpful. In this book the terms "engineer" and "designer" are used to typify several groups of professionals, including laboratory researchers, architects, software programmers, and all types of engineers, from mechanical to biomedical. "Engineer" can also refer to financial or actuarial professionals engaged in developing risk criteria for loan, investment, or business development purposes. The designer is a person engaged in the primary design of a system, component, or process. So, a typical designer may be an industrial designer, working on the look and function of a new toaster or drill for Black & Decker, or a designer may be a chef working on a new muffin for Panera Bread that does not crumble when eaten in the car. Similarly, the designer may be a financial portfolio manager working on a new method for managing clients of midsized companies.

It is my hope that each professional or group will be able to apply the principles as they apply to each specific situation. Within these chapters are thought-provoking, applicable, value-enhancing methods that help instill innovation in every part of an organization, ultimately increasing value and performance.

2

TURBINES OF SUCCESS

"Good ideas are not adopted automatically. They must be driven into practice with courageous patience.

—***Hyman Rickover***, *U.S. (Polish-born) admiral, 1900–1986*

Innovation can and should be found throughout the life cycle of new product development, from concept exploration through design and delivery. Even in the afterlife of product support, warranty, and ultimate disposal, there are opportunities to make more money and increase public and market goodwill. Opportunities to excel, to beat your competition, to gain customers, to increase positive public perception, and ultimately to increase valuation occur at every phase and are often hidden in the hinterland between the phases. We discuss the concept of product development life cycles further in Chapter 14, "Solving the Risk vs. Innovation Dilemma," and a basic definition is provided here to set up the rest of the discussion.

Business life cycles are broader than product life cycles. Business life cycles often include multiple product life cycles, organized in a sort of portfolio, with new entrants at their infantile beginning on the left and mature cash cows nearing the end of their profitable runs on the right. A level above the portfolio view is the overall organizational life cycle view. Here we see businesses in the development, growth, plateau, and decline phases. Understanding how all of these phases interact and how to optimize their contributions toward keeping the organization in the prized "growth" stage, just ahead of plateau, is an important concept.

On the surface the focus on product development phases appears to apply only to physical products. However, service industry processes go through the same phases, from concept inception to delivery, and are subject to similar market forces, timing, strategy, and delivery complexities. They may do it without a mechanical or electrical engineer, but they are otherwise similar in many ways.

Several books and references are available that explain in great detail the phases of new product development. Each industry uses slightly different definitions; however, the American Society of Mechanical Engineers (ASME, www.asme.org) has perhaps the most universally applicable. The aerospace and defense industry relies on similar definitions, and because of their quasi-governmental linkage, they are perhaps the most documented. Readers are encouraged to go to the International Council on Systems Engineering (INCOSE) Web site (www.incose.org) or to consult a book such as *Systems Engineering Management* by Benjamin S. Blanchard (Wiley, 2003) for more information on the basic phases. Blanchard's book uses basic definitions with the broadest applicability. Mr. Mike Bramley, senior software engineer for the General Electric Company, provided a good overview of the common terminology used in industry today:

> *The product life cycle consists of the following steps in which the requirements of the previous steps flow down until they are implemented in the product.*
>
> - *Product proposal*
> - *Marketing req.*
> - *System req. spec.*
> - *System design req. spec.*
> - *Component design req. spec.*
> - *Component design*
> - *Implementation*
> - *Verification demonstrates system meets requirements, link back to design requirements for system optimization*
> - *User feedback*
> - *Validation demonstrates system meets customer needs/intention*
>
> *After each major step there is a milestone review where improvements are made prior to moving to the next step.*

The key difference between the views presented in this text and the traditional views is the additional blending of the business considerations necessary at each phase. This blending is one of the keys to success found during the innovation blockbuster study. *The firms that understood the interdependency of the disciplines had a greater chance of capitalizing on innovation than those that were more compartmentalized.* This is no surprise, especially after the systems integration, enterprise resource planning (ERP), and business process reengineering crazes of the 1980s and 1990s. Interdependency of the disciplines is examined in this text, and the link to the capitalization of innovative products, services, and their development is demonstrated. The delineations between the phases of product development are mapped to the phases of business development, providing a unique model: the Innovation Lifecycle.

UNDERSTANDING THE BASICS

It is a gross oversimplification to provide a single picture of the phases of new product development and the innovation required in each phase. However, for purposes of discussion, some basic outlines of the steps involved with new product development are given. A link can then be made to the technical, business, and strategic aspects of product development. A basic picture of the phases is shown in Figures 2-1 to 2-3.

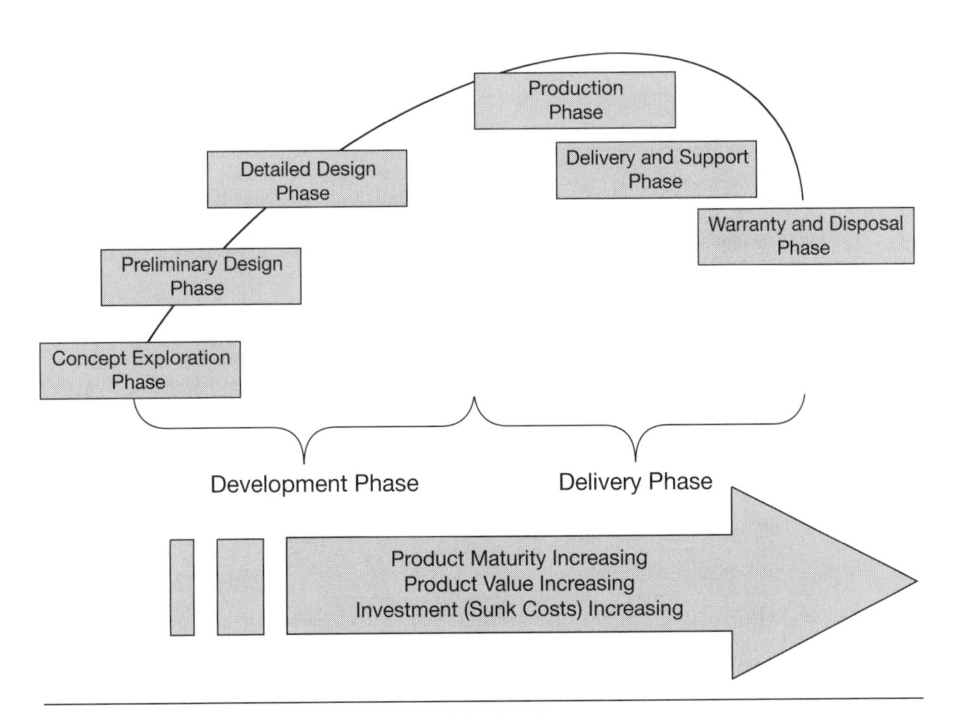

Figure 2-1 Basic product development and design phases.

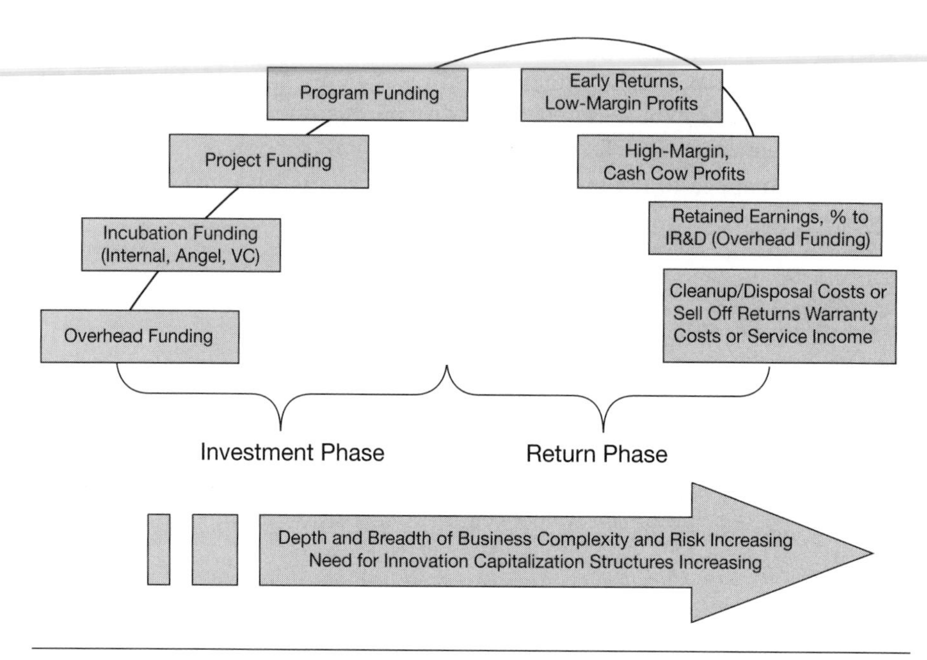

Figure 2-2 Basic funding and return phases.

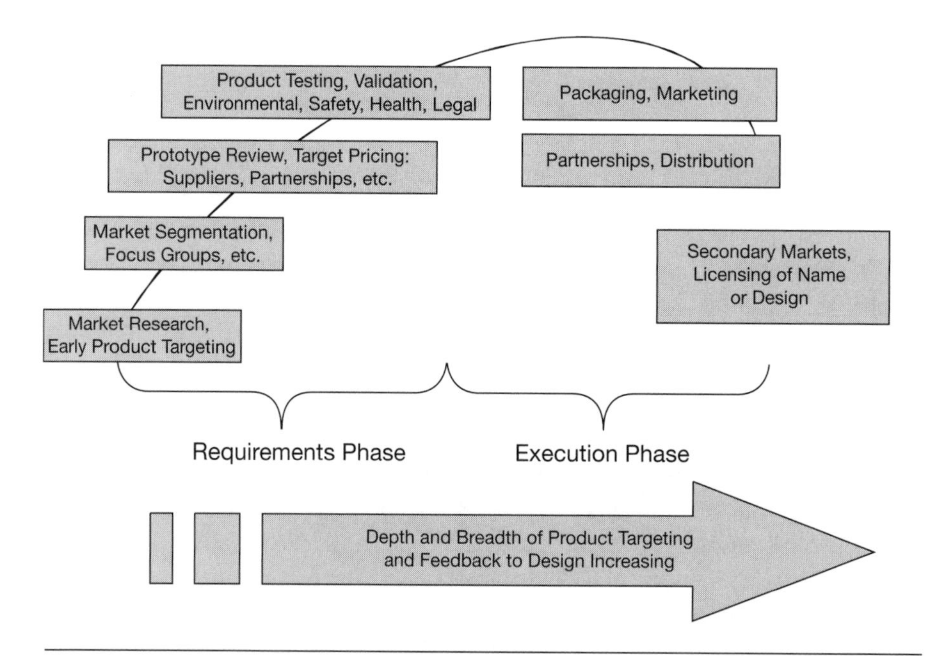

Figure 2-3 Basic marketing and strategic phases.

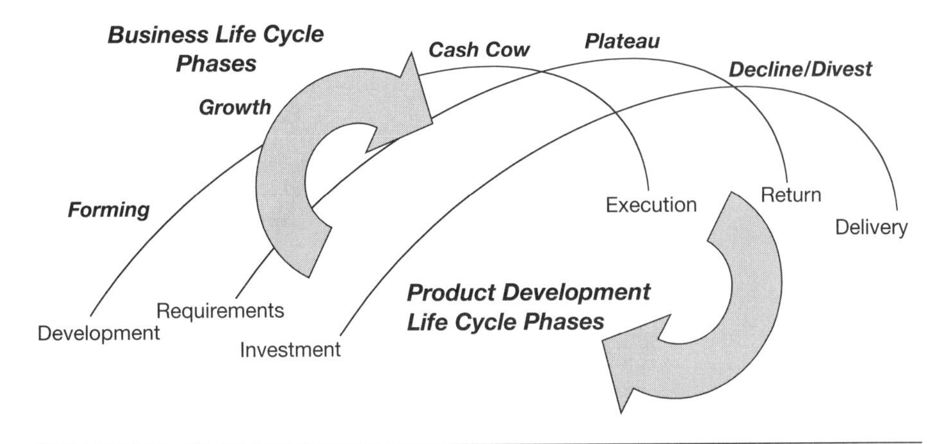

Figure 2-4 Assembling the construct. Product development and business phases are intertwined and interdependent.

THE INNOVATION LIFECYCLE: COMBINING NEW PRODUCT DEVELOPMENT PHASES AND BUSINESS PROCESSES IN A SINGLE PICTURE

Now we overlay an innovative picture on top of the basic development phases in the earlier pictorials. This forms a construct or map for the innovative life cycle process. Providing a clear, easy-to-remember physical embodiment of the business and technical engineering aspects of the phases was the main goal of this map, a skeleton that roughly embodies all of the nuances in new product development and also includes the results from our study on innovative blockbusters.

Overlaying the three previous graphics, we start to see a sort of repetitive rhythm but with an unmemorable graphic, shown in Figure 2-4. If we then decouple and couple the activities within each major phase, we see that they are intertwined in a spiral fashion, where there is innovation within an activity and innovation within a linkage between activities. See Figure 2-5.

The venerable turbine engine satisfied this criterion and provides a composite of Figures 2-4 and 2-5. The turbine engine can be used to help one understand the fundamental elements of product development and the phase of innovation. When the turbine engine is superimposed on top of the basic development phases shown in Figures 2-1 through 2-5, a construct or map for the innovative life cycle process emerges, as shown in Figure 2-6. The goal of this mapping is to provide a clear, easy-to-remember physical embodiment of the business and technical engineering aspects of the phases. The turbine engine representation is

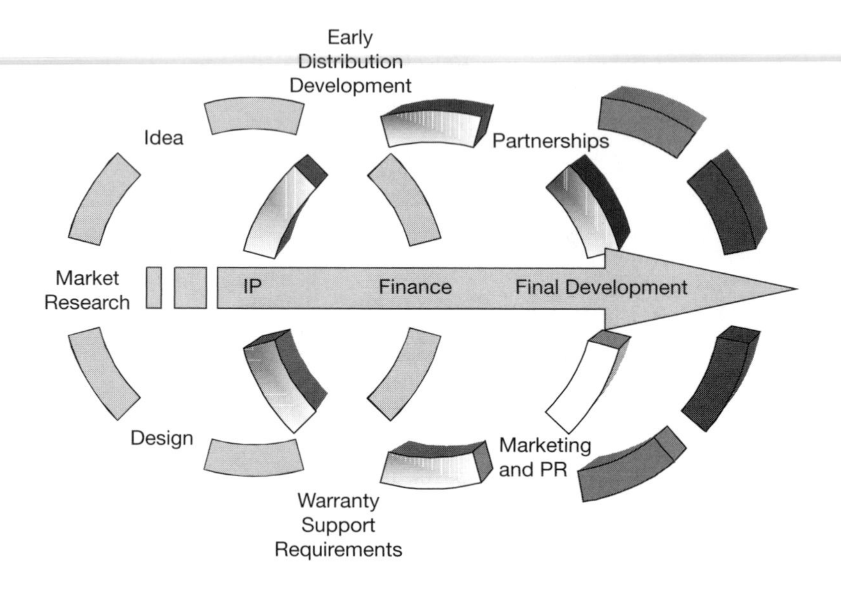

Figure 2-5 Overlay of development phases and business cycles: a spiral view of risk and business phases.

used throughout the remainder of the text to depict the full process and nuances of new product development.

Why the turbine engine? Why not some other engine or some other device or natural process? Simple. The turbine engine typifies the power associated with innovation. The turbine engine has the highest thrust-to-weight ratio of any power plant, including nuclear. (OK, the mighty ant's amazing anatomy eclipses the turbine, but I do not understand enough about ants to use them as a framework for innovation.) Only the elegant Stirling engine is more efficient, close to being a perpetual-motion machine. But the Stirling cannot move a 300-ton airliner through the sky at 500 miles per hour. Only the jet turbine can.

UNDERSTANDING THE TURBINE

So, let's take a closer look into this marvel of machinery and fossil fuel. "Intake—compress—combust—thrust." Those were the profound words of my college aerospace engineering professor on the first day of class. I still remember his Eastern European accent and the deadpan way he described each stage of the turbine. He used "intake—compress—combust—thrust" to describe the basic workings of the turbine engine. The advent of the turbine engine made many things possible that were merely pipe dreams just years before, such as trans-Atlantic air

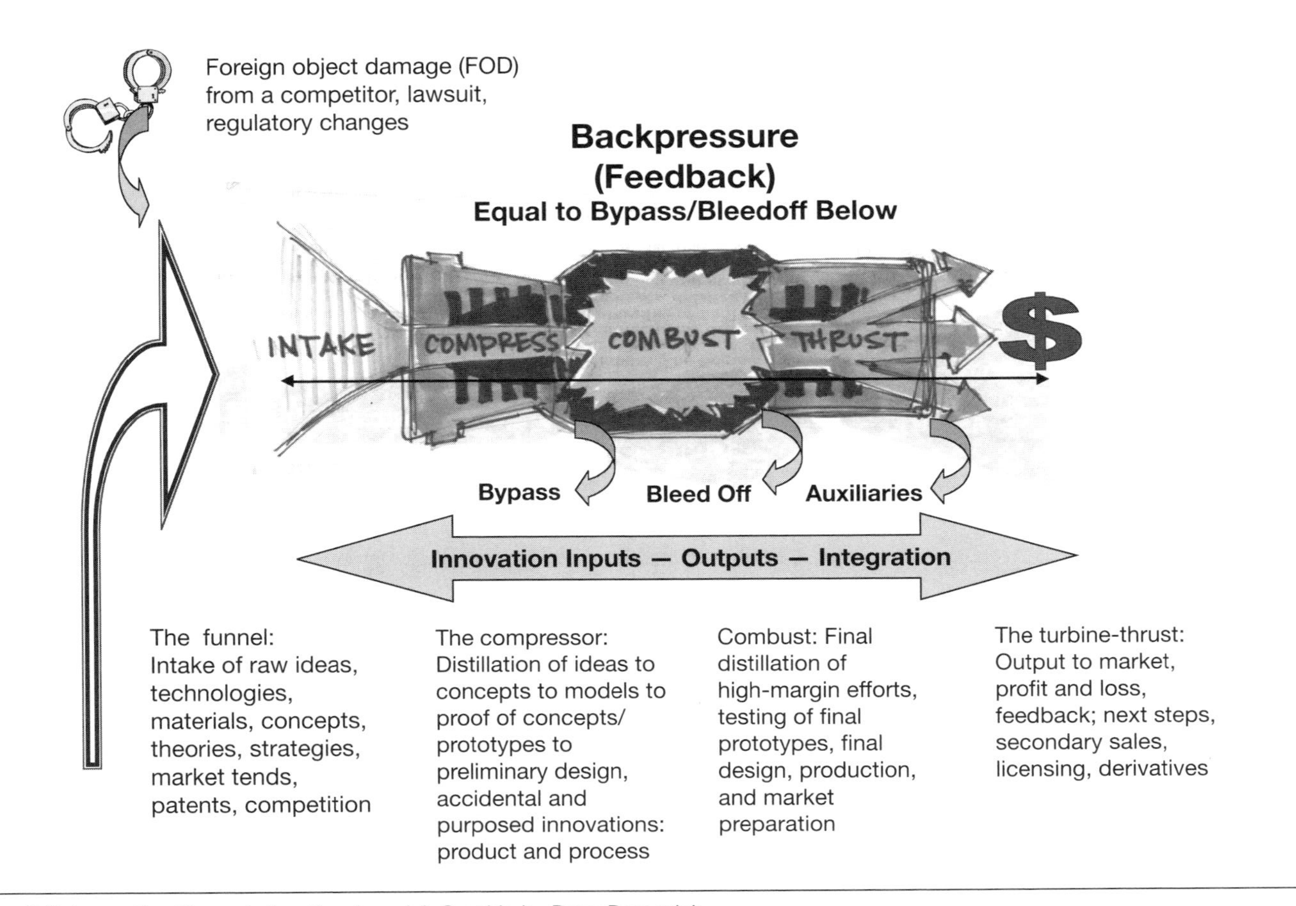

Figure 2-6 Innovation life cycle functional model. Graphic by Drew Dernavich.

travel, global delivery of goods and services, faster and better war-fighting machines, and in many parts of the world, electric power. Chrysler even made a turbine-powered car in the 1960s that recorded amazing performance and fuel economy rivaling today's economy cars during a cross-country test.

As I write this, coincidentally, I am flying over the Atlantic (ironically passing over Combustor, Maine) on the way to London, England, in a Boeing 777 doing 616 mph ground speed—all on two General Electric high-bypass turbine engines. While turbine theory predates the 1950s embodiment of the first commercially available engines by many decades, it was innovation in materials science and computer science that finally made them practical and reliable. On the materials side, the metals and casting techniques necessary to withstand the high internal temperatures and stresses were not available until the 1940s and are still under development and refinement today. Computers that assisted in the complex computations necessary to prevent air stall on the blades and vanes inside the engine greatly assisted the development, as did countless other innovations in the areas of welding, mathematics, computer numerical control (CNC) machining, and so on—all driven by the massive economic drivers linked to the ever improving thrust-to-weight and thrust-to-fuel ratios of the turbine.

Intake—compress—combust—thrust. The phrase has a nice flow to it—a built-in harmony of activity. This harmony of activity precisely represents the flow and stages found when one is looking at the basic product development process and the success stories that made the business world take notice. It was the model that held true for every basic science, business, and management discipline I looked at in the preparation of this book. The turbine engine model is able to include many topics that are covered in this book, from a basic "product development with consistent innovation" discussion to other success factors such as personnel policies, finance and accounting policies, design processes, management processes and practices, even cash flow and portfolio management.

That turbine model combined with the common outputs of each stage is shown in Figure 2-7.

Intelligent Innovation seeks to prove, encourage, explain, and provide instruction relating to three basic principles:

1. Innovation must be pervasive for blockbuster success. Innovation must be found, encouraged, used, and capitalized on in every phase of development in a firm, from the early concept phase through delivery and support. In short, a life cycle approach to innovation is presented.
2. Innovation is necessary for corporate stock growth, regardless of all other influence factors. Corporate growth cannot occur without consistent innovation in product development.

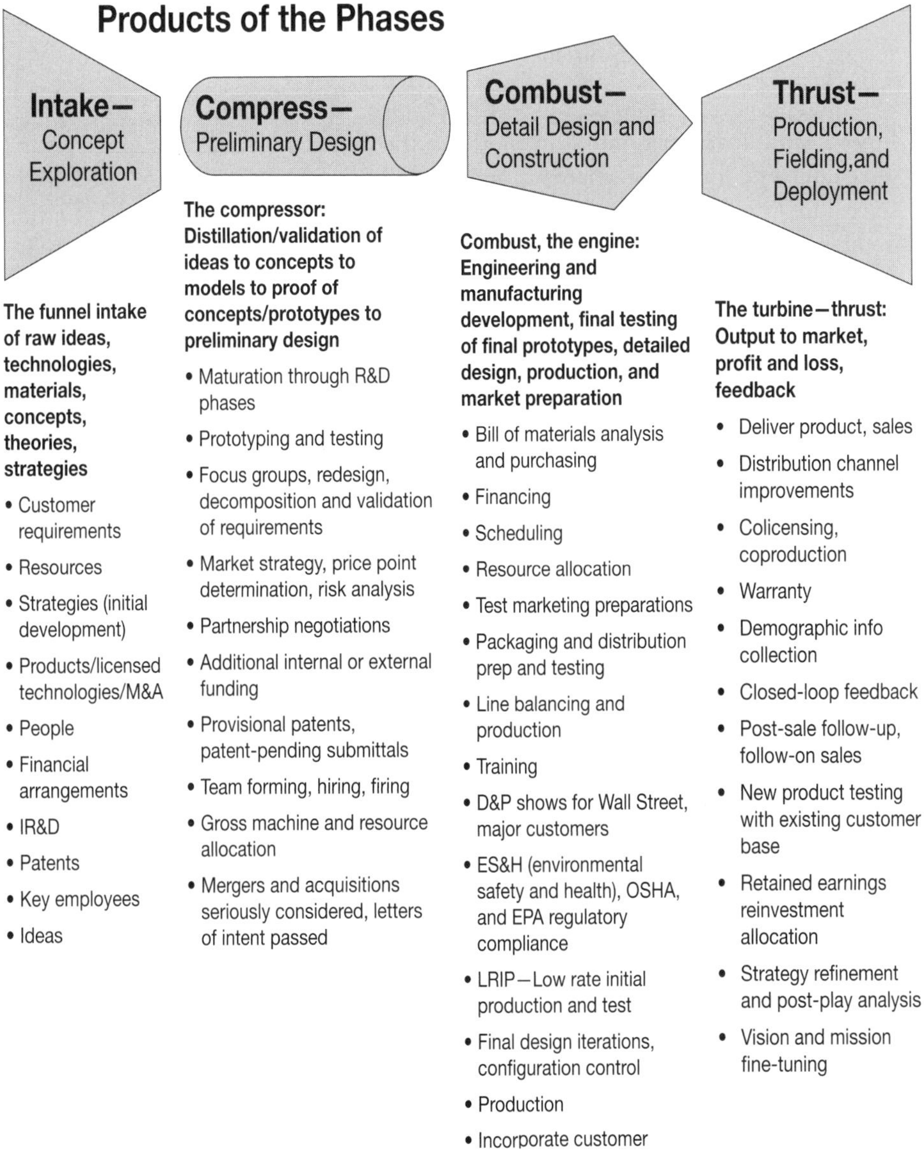

Products of the Phases

Intake— Concept Exploration

The funnel intake of raw ideas, technologies, materials, concepts, theories, strategies

- Customer requirements
- Resources
- Strategies (initial development)
- Products/licensed technologies/M&A
- People
- Financial arrangements
- IR&D
- Patents
- Key employees
- Ideas

Compress— Preliminary Design

The compressor: Distillation/validation of ideas to concepts to models to proof of concepts/prototypes to preliminary design

- Maturation through R&D phases
- Prototyping and testing
- Focus groups, redesign, decomposition and validation of requirements
- Market strategy, price point determination, risk analysis
- Partnership negotiations
- Additional internal or external funding
- Provisional patents, patent-pending submittals
- Team forming, hiring, firing
- Gross machine and resource allocation
- Mergers and acquisitions seriously considered, letters of intent passed

Combust— Detail Design and Construction

Combust, the engine: Engineering and manufacturing development, final testing of final prototypes, detailed design, production, and market preparation

- Bill of materials analysis and purchasing
- Financing
- Scheduling
- Resource allocation
- Test marketing preparations
- Packaging and distribution prep and testing
- Line balancing and production
- Training
- D&P shows for Wall Street, major customers
- ES&H (environmental safety and health), OSHA, and EPA regulatory compliance
- LRIP—Low rate initial production and test
- Final design iterations, configuration control
- Production
- Incorporate customer

Thrust— Production, Fielding,and Deployment

The turbine—thrust: Output to market, profit and loss, feedback

- Deliver product, sales
- Distribution channel improvements
- Colicensing, coproduction
- Warranty
- Demographic info collection
- Closed-loop feedback
- Post-sale follow-up, follow-on sales
- New product testing with existing customer base
- Retained earnings reinvestment allocation
- Strategy refinement and post-play analysis
- Vision and mission fine-tuning

Figure 2-7 Turbine phase flowchart with typical outputs/activities by phase.

3. Innovation transcends traditional boundaries; it must be in product, process, and people. Innovation in one area without innovation in another is suboptimal.

Intake—compress—combust—thrust are the actions of Intelligent Innovation transformed into four modules: the funnel or intake, the compressor, the engine or combustion chamber, and the turbine. Some planes also have a fifth stage built into the turbine: the afterburner. The afterburner is typically only found on high-performance, agile military aircraft such as fighter jets. Its sole purpose is to provide raw, additional extraordinary thrust (with a corresponding raw, extraordinary consumption of fuel). Likewise, only a few firms have the processes that equate to the afterburner, and I will be discussing that in a separate book.

Each of the four main sections of the turbine are discussed in corresponding chapters. Each section, function, and subfunction of the engine example is critical to the success of your organization. Although product-oriented firms are mentioned during the explanation of the engine analogy, any organization, for-profit or not-for-profit, or service can use the model.

In the turbine, each module is fed by the previous one and/or by the external environment and is controlled via feedback loops to the engine control unit (ECU; in other words, process management). The feedback loops in an engine are critical to monitor things like internal pressure, shaft speed, and fuel consumption. Likewise, feedback loops in a corporation or any organizational group are important. In the context of Intelligent Innovation, such loops provide critical linkages between the phases of product development, between product pricing and materials selection, between competitive analysis and white space targeting, between risk and return, and ultimately between strategic intent, action, and results.

THE FOUR PHASES IN DETAIL

Intake: The Funnel

The first stage of the product development process, also known as the funnel, is the most celebrated and documented phase of development. This part of the engine correlates to the generation and assimilation of innovative concepts. Much of the documented work in the intake stage is naively compartmentalized. To look at the first module of the turbine without the context of the other modules is to produce highly volatile rocket fuel with no place to store it, no plans to utilize it, and no ability to control it for good. Indeed, the production of rocket fuel is in itself an expensive and time-consuming endeavor, and to do so without ability to capitalize on it is a waste.

This stage is commonly referred to as the "intake" of the engine. Air molecules are drawn into this stage of the jet engine by the vacuum created by the rotating first stage blades.

During this initial stage in new product development, ask yourself these questions:

- Does your firm create a "vacuum" or constant pull for new ideas, meaning is it open to such ideas and to innovative thinking?
- Are the new ideas revolutionary? Do they add significant value, lower costs by a large margin, or increase performance over the current state of the art?
- Does your church or school create a pull for new members? Is it creating new programs that meet the current clientele, or just polishing the same old programs that were designed for a world 40 years ago?
- Does your firm create a vacuum to attract new partnerships, technologies, or customers? Can it adapt to new business models—beyond the addition of new processes and products?

Are people clamoring to get into your town, or are they leaving in droves? What is the reason, good or bad? Is it the taxes, the trash removal, the schools, the roads, the integrity of government? These are the defining forces of innovation.

Why has so much been written about this first phase? Well, it is loads of fun. The conceptualization of new and outlandish ideas and the blank slate of business creativity is a blast. We all imagine patents in our name for widgets that save the world and make us millionaires. Indeed, this module also includes the "purchase" of creative input, as do all the other modules. The funnel contains ideas, patents, white papers, studies, sketches, discussions, competitor products, competitor information, primary technologies, creative and high-risk funding mechanisms, and just about anything else that can be construed as a building material for the output of an organization. While it is easiest to use hard product examples—for example, a new miniature hard drive that makes handheld computers more viable—it is important that nonproduct organizations realize that this module is just as critical—even more critical, in fact—for their success.

One of the key reasons the funnel must be integrated into the compressor, engine, and afterburner is growth. Growth cannot be obtained by or through sporadic new product development. The underperforming firms in the SRM study had often made a significant push and corresponding investment in creative front-end-of-the-business (FEOTB) endeavors only to stop the effort after an (insufficient) amount of time. One leader's push for innovation and creativity was replaced by another leader's push for "core competency"—that is, getting back to basics. My guess is that had the first leader's push for FEOTB activities been thoroughly

CASE STUDY: STUBBLEBINE COMPANY

David Stubblebine runs a very successful regional commercial real-estate firm. It relies on various phases, some quite advanced, to bring business in, work out features of the deal, close the deal, and record the results as profit. The funnel for the Stubblebine Company is real-estate listings. There are no high-tech patents to license and figure out, no complex multinational merger to negotiate before the engine can begin to spin, and no cutting-edge product that must be rushed to market to gain the lead. In real estate it all starts with the listing. While the rest of the commercial real-estate business process is quite complex and can involve build-to-suit negotiations, complex financing vehicles, and knowledge of endless laws and regulations, this early stage of development, getting listings, is more elemental by comparison. The Stubblebine Company must make sure the listings "prime the pump."

Remember the old 1980s commercial for Dunkin' Donuts, where the tired-looking man would get up at 4 a.m., mumbling, "Time to make the donuts . . . time to make the donuts" or there would be none on the shelf for his customers? He had to make sure the donuts would be fresh and appealing to customers. Similarly, a real-estate agent is only as good as the listings for sale. These listings represent buildings that are empty or soon to be empty. They are listings that the next agent could get instead. These are the same listings that have been on and off the same market for 20 years.

Although his father had built the Stubblebine Company into a very respected, small regional firm over a period of 20 years, when Dave inherited it, it only had a few listings and five employees. David was concerned about that ratio and applied several creative proprietary techniques and some organizational skills to turn that ratio around.

But how does the "Intelligent Innovation funnel" apply in this case? What happens inside the funnel for this part of the process? In short, everything—everything is at stake in the funnel for Dave. Because three listings for five agents didn't make sense, he set out to "fill the funnel." Today the company has 100 listings for those same five employees, plus those available on the Multiple Listing Service (MLS) system.

Now Dave sees considerably more profits than five years ago. The number three agent at the company closed on six listings in the week before Christmas. This represents over 40 percent of that agent's take for the year and a considerable percentage of the firm's take for the year. All six of those sales came from listings that were scraped together the year before, with low probabilities of closure. The agent and the rest of the firm worked hard for many months. Success is more than a listing however. Those listings were closed with the help of finely tuned advanced back-office processes, understanding of the legal and financial aspects of the deals, solid ethics, and negotiation skills of the entire firm. The funnel and the rest of the processes sustain the life of this firm.

integrated into the fabric and process of the rest of the firm, it would have been a rousing success. While each stage does integrate into the overall process, the first is constantly seen as separate in the popular business press—as though it was something you dust off every five years, perform, and then get back to business. In fact, the blockbuster firms have this first stage tightly coupled to resource allocation, R&D budgeting, strategic planning, competitor analysis, and so on.

The jet engine is nothing without massive amounts of air and fuel to be compressed, ignited, and thrust out the back. However, the air must enter in exactly the correct ratio for the engine to operate efficiently, either by design of the inlet or by controls placed on the inlet. Every aspect of the organization starts at the intake stage, from talent acquisition to internal research and development.

Compress: The Compressor

This critical stage is where air is compressed an amazing 10 to 30 times its original density and pressure. It corresponds to the place where new ideas go to be trimmed down or beefed up after they are conceived. As new product ideas are allowed to go through the intake to the compressor stage, they become denser, the value added to them increases dramatically as prototypes move into more mature stages, and proof of concepts indeed prove or disprove concepts.

Ask yourself these questions:

- Are your ideas maturing at the right rate for your organizational goals? Are you capitalizing on the ideas and investments at the right rate?
- Are you developing enough ideas and the right ideas for the marketplace you are in? For your investors and customers? For now and in the future?
- Are your employees feeling satisfied or frustrated? Do they feel held back by a clunky process or empowered by a can-do, smart process with adequate checks and balances?

I have a sort of odd hobby that is related to all of the stages but particularly this stage. I frequent discount stores all around the globe. You know, those stores in the "second tier" of retailing that sell "as seen on TV" products from the year before—products that Sears, Wal-Mart, and other major retailers either discarded outright or gave up on. Those products tell an amazing story in their relative failures. While I acknowledge that some make a stand in this second tier and perhaps eke out a profit, most are there because they have been deeply discounted and are no longer in production. Those firms got the compression cycle wrong. They let too much through the compressor, and they let products through that were not sufficiently compressed, tested, refined, targeted, and priced.

Another sort of ratio mismatch can occur when too many new product ideas are let into the compressor and combustor for the system to handle and the entire system stalls. In effect, the volume of new products "blows out" the flame in the later stage. In a real engine, this is called a "rich" mixture—that is, too much fuel. However, the first example—underrefined offerings—is more common. In a real engine, this is called a "lean" mixture—in other words, too much air. This is perhaps the most complex part of the process, the part where many firms and organizations fall down. It is the area of the engine where the life-giving, thrust-producing air is compressed further with just the right ratios and speed. It is the least understood phase by the business schools; many do not even acknowledge its existence, and literally hundreds of products have failed at the hand of high-priced workers with prestigious master of business administration degrees. For those of us in the R&D environment, it is the most exhilarating, rewarding, and challenging job on the planet: bringing napkin sketches through the prototype, test, preliminary design, and detailed design phases.

Combust: The Engine or Combustion Chamber

What does one say about an area of the engine that reaches beyond the melting point of exotic high-alloy metals? A place where air rushing at hundreds of miles per hour is accelerated to thousands of miles per hour? This is where a flame must burn constantly amid the swirling, exploding, accelerating pressure of air and fuel. In short, this is where it all happens. The process correlates to the production floor, where risk, strategy, finance, and a host of other ethereal discussions all take a backseat to blood, sweat, and the schedule. The schedule is king here, where even a small hit on performance or quality can be absorbed, corrected later, discounted, or passed on to the consumer, as long as the schedule is kept. It is the production floor where only the truly brave make their living—where tempers flare, statistics cause pressure, stocks react, and the product is shipped.

In a real jet engine, the compressed high-pressure air rushes in from the compressor section and blows over a "can" where the igniters are. Fuel is sprayed by fuel injectors into the can, mixed with the air, and ignited in a continual rush of energy. The can helps prevent the fire from blowing out and helps distribute the flame, or creates a controlled explosion. In a traditional product development scenario, combustion is the epitome of the value-added chain. It includes the final design from engineering, all of the checks and balances to ensure the design is ready, and all of the components and skills that are assembled to realize that design. It is everything to which form and transferable value are given, culminating in either a product or a service that is deliverable to the market. It is not delivered yet, however. That comes next.

Thrust: The Turbine

Thrust is where fortunes are made. The turbine is the set of blades and vanes that are directly coupled to the same shaft as the compressor at the end of the engine. The hot, explosive, compressed gases coming out of the engine stage rush past the turbine stage out the back as thrust—or, in the analogy, profits. Part of that energy is returned via the turbine blades and shaft to the front of the engine to keep the cycle going—retained earnings in the analogy. The turbine correlates to the point in the process when revenue is actually realized. It is controlled by corporate policies, ethics, and generally accepted accounting principle (GAAP) rules. It extracts energy from the firm and delivers it to the market at a profit. Any measure of "margin" is equal to thrust. Zero margin equals zero thrust. This means the fuel was wasted in the previous stages, equal to cost. A healthy process yields a little extra, called retained earnings. The retained earnings are used to rebuild aging equipment, train employees, give holiday bonuses, put in a rainy-day fund, pay dividends, and the like. In the engine, this is called reserve thrust.

The afterburner is a crude device found primarily on high-performance military engines—mostly fighters, which need a thrust-to-weight ratio far exceeding what the plane needs for level flight. It consists of a series of fuel nozzles that spray fuel into the volatile high-speed mix of air coming from the turbine. This fuel is ignited for a "wiz–combust–wow" addition of raw thrust. While that thrust does not come cheap, its return can be enormous, enabling fighter planes to accomplish amazing aerobatic moves and speed. The afterburner is analogous to licensing a firm's primary or ancillary technologies or know-how. The firm makes a return on the technology through its normal delivery channels and then another return on the primary technology by licensing, even to a competitor. This can comprise secondary market sales whereby products or services are sold to foreign countries with little or no additional investment or development. This component is essentially a derivative margin-producing mechanism that is often overlooked. It is possible to try and "buy" the thrust stage by outsourcing all the rest. Remember the dot-com craze, now called the dot-bomb daze? Dot-coms promised to reshape the landscape, to provide goods and services with little or no delivery cost, to create wealth by doing nothing except connecting the previous four stages to the World Wide Web. Well, it was soon discovered that customers want to buy from real companies attached to the previous four stages. The dot-coms found out quickly that they needed to actually provide a product and that it was a bit naïve to think a firm that worked on the first three stages would simply hand over the product to the fourth stage without requiring sufficient margin to make the first three work.

The delivery of goods and collecting revenue is not enough. There must be service after the sale, and service requires people, training, systems, honesty, and money. Essentially, there is no pure play for the fourth stage; even Las Vegas gambling businesses think of their customers in the same way as Lexus or IBM. They engage in marketing with premiums and incentives (the intake); they sign people up for special weekend packages (the compressor); they deliver a fun, entertaining package full of food, shows, and gambling (the combustion chamber); and they expect and realize a return of gambling profits (the thrust). All of this is controlled at exactly the right rate to deliver a fun-per-dollar ratio that keeps the customer coming back and the casino in the black. This carefully measured give-and-take of the casino is analogous to the ECU and feedback.

CHAPTER REVIEW

In this chapter we reviewed the basic building blocks of the Innovation Lifecycle. The steps required include the orchestration of the following:

- Requirements and execution
- Investment and return
- Development and delivery

Since these phases of development are intertwined and interdependent, they must be linked. The turbine engine's central shaft provides an effective linkage between the furthest phases of intake and thrust, and likewise an organization's policies and procedures link requirements and investment all the way to execution and delivery. We explore the management and control functions necessary to optimize these complex events in the next chapter, and we also explore the criticality of policies and procedures, which can make or break any effort in any phase throughout the book.

3

ENGINE CONTROLS

> "Technology is dominated by two types of people: those who understand what they do not manage and those who manage what they do not understand."
>
> —***Unknown source***

The SRM Survey revealed a number of common factors that led to successful introduction of new, innovative products. These factors, subsequently referred to as I-Factors, were found across industries and throughout phases and were insensitive to the size of the project or firm. This chapter discusses the first two I-Factors and their relationship to innovation, with a focus on decision management (DM), one of the key contributors to success across industries and phases. Other I-Factors are discussed in subsequent chapters. For easy reference, they have also been compiled in Appendix F.

The SRM Survey data provide conclusions in three major segments:

1. Decision-making efficiency
2. Innovation capitalization (linked to corporate policy)
3. Strategic risk/return understanding.

And so we focus on the first I-Factor in this chapter:

> **I-Factor 1: Any success, personal or business, is rooted in good decision making. Innovative decision making separates the wildly successful from the rest.**

For example, the SRM Survey data show a reasonably strong link between solid business performance in the aerospace and defense (A&D) industry and the use of systems engineering requirements management tools and methods. The A&D industry is heavily dependent on a rigorous engineering method called systems engineering for requirements and system design definition. Systems engineering is geared toward the earlier phases, where extensive customer interaction produces a thorough, written set of project requirements. These requirements are used throughout the project to create trade studies, develop contractual milestones, choose materials, analyze performance levels, and so on. Essentially, the requirements are used to make decisions for everything from design to test. They are used as a type of control on the design process.

The airplane seat was getting uncomfortable. I needed a walk, but the passenger in the aisle seat sitting next to me was sound asleep. Hmm, I mused. Does a 7 percent stronger link to requirements equate to a better product brought to market faster with more satisfied customers and less redesign and rework? Hmm, I mused again. This is clearly a leading question that the lady did not care about in her slumber. And what of this supposed conflict of hers. Is innovation more important than engineering and management? I cannot separate them in my head. Anyway, back to my requirements linkage. Perhaps an answer is there. Let's explore this lead on a napkin. If a stronger link to market, customer, or organizational requirements makes the resulting decisions more "accurate" or, perhaps more importantly, more "relevant," then those decisions must be "better." Better means more efficient, requiring less rework. It then follows that the benefit of this newfound wisdom could be realized in employee hours saved through less rework or though increased sales.

I smile at her. Maybe the numbers will awaken her. After all, any spark of "innovation" she holds so dear can only reach its intended audience if it is acted upon, if it takes real form and is delivered. That requires decisions, lots of decisions. If those decisions are more efficient, the innovation delivery will happen, sooner and better.

GOOD DECISION-MAKING PROCESSES SAVE TIME AND MONEY

This concept, also known as decision management (DM), correlates to the ECU of the engine. The ECU controls air pressure, shaft speed, bleed air, cooling temperature, and fuel consumption, ultimately providing thrust reliably, hour after hour, day after day. DM is the overall controlling linkage in the organization; it is the orchestra leader. If a turbine engine had a faulty controller, it would easily overspeed and break apart from the forces, or it would bog down with too rich a fuel mixture, belching smoke instead of thrust. Likewise, firms with established DM programs exhibit a 7 percent better link to requirements basis and an astounding 91 percent greater probability of having a clear and clearly disseminated vision and mission statement. Subsequently, there is greater clarity and direction for people throughout the organization. DM is part of each phase of the product development cycle. It cannot be separated out and should only be described from an integrated point of view. The concepts of vision and mission will be revisited in Chapter 8, "The Strategic Balancing Method" and Chapter 16, "Innovation in Culture and Attitude: Revving the Engine."

DECISION MANAGEMENT MUST BE INTEGRATED WITH THE WHOLE LIFE CYCLE

To make a decision in the intake phase—such as which firms to buy, which technologies to invest in, and which ideas to run with—without understanding how that decision is dependent on the compress phase and later the combust phase is a drastic mistake. Notice I used the word "dependent" and not some more popular term such as "linked" or "integrated." Some 20 years ago I wrote a white paper on the integrated nature of decision making and espoused how important "integration" was in all aspects of the organization. The paper was subsequently published in the proceedings of an engineering symposium, and I presented it a few times. Unfortunately, years later I realized that I was slightly off target. While it is true we need to integrate all aspects of our organizations, it is also true that most parts are dependent on one another and even dependent on exterior forces, customers—or partners, if you take a life cycle view. I talked a lot about integration but missed the complexity of the dependencies that it creates.

This is an unpopular view. It is nearly impossible to teach, or so say the schools. It is nearly impossible to write about, or so say many popular business and engineering texts. It is unpleasant to accommodate, or so say the personnel directors. It is very expensive to support, or so say the ERP implementers. Unless there is both an action and reaction for each decision and the internal and external forces act on that decision in an integrated and dependent model, the outcome will be, at best, suboptimal because of gaps in delivery and, at worst, a failure. This holds true always in the long term and frequently in the short term.

CASE STUDY: DECISION MANAGEMENT WITH FIAT

When GM decided to buy a significant chunk of Fiat in 2001, the purchase looked like a good idea. Fiat was devalued. It had developed some leading automotive technologies and had good manufacturing and design abilities. It had some of the most innovative assembly and interior technologies in the automotive industry. Arguably, it had overinnovated. Innovation and idea generation can be like a drug, and designers get addicted to a good thing.

Overinnovation can occur anywhere at any time; one has simply to walk through a modern full-size supermarket chain store. The plethora of niche and me-too products on the shelves is staggering. One can easily count 20 types of plastic wrap for preserving food or 20 types of instant noodle soups. Snack foods top the list, with over 100 distinct types in most markets, all scrambling for a share of the $6 billion annual U.S. market. Can each one produce a profit to justify all of the costs, including the hefty shelf placement costs? While the rate and quality of new product introductions, particularly disposables, in the home-based supermarket segment are astounding, they do border on overinnovation.

Fiat's overinnovation and choppy market segmentation could be fixed, according to prevailing logic. Fiat had a good reputation and distribution network in Europe. However, GM failed to see the long-term implication of the buy. Perhaps it failed to look closely enough at failure modes and the integrated nature of politics and business in Europe. In my opinion, it missed several major dependencies. When Fiat's market share and financial troubles worsened, GM had a choice: Stick with Fiat and go down with the ship or pay a huge ransom and save both ships but end up in a severely compromised position. According to a BBC News article of Monday, February 14, 2005: "General Motors of the U.S. is to pay Fiat 1.55 billion euros [$2 billion, or £1.1 billion] to get out of a deal which could have forced it to buy the Italian car maker outright."

Having opted for the latter, GM had a huge payout, which came at the worst possible time. Its competitors were finally introducing new models and taking market share. Its gas-hungry but profitable truck sales had stalled because of high gas prices. A good gap analysis with dependencies may have predicted this as a possible scenario. Fiat began to fail for many reasons, some similar to its huge partner, and some quite different, but all quite predictable. Fiat was faced with high labor costs, a confusing overoptioned product offering, lack of solid partnerships, and mounting European and Asian competition. GM either ignored or overestimated fixing these issues when it entered the deal, and now it was stuck with an unhealthy dependency where a good one could have existed.

In the end, Fiat seems to have won. According to the article, Fiat's CEO Sergio Marchionne said: "We now have absolute freedom to design our own future." Clearly, he did not like the idea of dependencies, even good ones.

A corollary to I-Factor 1 emerges: the willingness to look both inside and outside for dependencies increases decision accuracy. Finding and exploiting DM dependencies is a key aspect of Intelligent Innovation. When UPS realized that its customers were dependent on the company, it took step one and analyzed its customers' requirements. It used these requirements to develop better service offerings. It used research on its customers to identify and meet its customers' needs without requiring any changes on the customers' side. UPS believed in partnering with its customers. This vision went beyond a basic service mentality and worked to give customers new tools to make the process more efficient and subsequently tied them to UPS systems. This in turn helped UPS. When UPS also realized that it was in a codependent relationship with its customers, real integration began to take place. UPS went beyond basic requirements and delved into its customer value chains, helping solve real problems and integrate UPS and other partners into the solutions. This extra step was the difference between survival and excelling in a very competitive industry.

Interdependent relationships between customers and business work for both service and manufacturing organizations. The once old, stodgy United States Postal service followed the example set by UPS. Now through the use of innovative systems from Electronic Data Systems (EDS), customer-oriented thinking, and partnerships with firms like PayPal, a customer can buy postage, apply postage, and ship packages without ever entering a post office. Just a few years ago that would have been ridiculous to suggest.

THE COST BENEFIT OF GOOD DECISION MAKING

The real benefit of DM is the savings associated with more efficient decisions. In this simple example using real rates from the A&D industry, assume the average decision (for a product or report, etc.) involves 10 people. Two at the lowest level, whom we will call junior engineers, are at a pay grade of 3. They collect data, go to various meetings to discuss it, write a paper, and alter drawings based on the data, for a total of two worker-months of labor. Then three people check their work, one literally and two in a more general sense in hallway discussions, for a total of one worker-week of labor at a pay grade of 5. Then three executive-level people utilize the information and make minor modifications, for a total of two worker-days at a pay grade of 8.

The modifications send the juniors back to the drawing board for one additional week, along with some grumbling. Finally, a clerical person records all the final information in various formats: two worker-days at pay grade 2. If we use typical rack rates for a midsized U.S.-based manufacturing and engineering firm—ranging from $190 per hour for the grade 8 to $53 fully loaded for the grade 2—this hypothetical design–report–product–decision cycle costs $29,000. Saving a scant 7 percent of that (the number from the study) by using a more robust, targeted, relevant, and integrated DM process saves an easy $2,000 per decision.

"Wow!" I exclaim again. The indifferent person next to me stirs for a moment. I say aloud, "One could argue that this improved DM ratio could be applied through the entire structure of a firm, and the savings, if annualized, are tremendous. A $1 billion firm with a payroll of, say, $400 million could save (or increase output) about $28 million, straight to the bottom line. And to think Dr. Evil would blow up the world for only $1 million. I can give him $28 million just by making more clearly communicated, more relevant (top-down and bottom-up) decisions. This does not even account for the potential increase in sales that would result because the products and services are more accurately targeted in their design and delivery."

"What does this have to do with innovation?" she asks. Hmm, she was not sleeping after all.

I get back to work, trying to maintain my confident expression.

Decisions are most effective when made by people with vision. Are people with a vision always "successful"? Patton and his 3rd Army, along with Rockefeller, Carnegie, Honda, and Ford, did pretty well. But then again, many other visionaries were unsuccessful. Remember Hughes, Mussolini, and the Enron executives? Clearly, more than just vision and some good decision making is needed for success.

The ESRM Survey clearly shows a link between the use of vision and mission success. Since 91 percent of those surveyed used a solid vision and DM, this requirement became one of the primary success factors for innovation. In an earlier work, I used a term called "decision fulcrums" (in an article entitled "Strategic Balancing: An Integrated Methodology for Management and Efficient Decision Making," *EIA/GEIA 2000 Engineering and Technical Management Symposium Proceedings,* September 2000), which communicated the interaction of certain aspects of decision making on project management. Decision fulcrums have a systemic positive effect on decision-making efficiency because they provide a decision point from which to get moving. More than a catalyst, they are an impetus, demanding action in any direction. A decision fulcrum—for example, a hard due date for an application or a contractual item due on a certain date at a certain level—mixed with a clear vision helps employees at all levels know where they are going and how their work fits into the whole process. The ESRM Survey is filled with anecdotal stories of firms that drove toward clear goals and succeeded despite numerous constraints and shortcomings. These stories appear to support this theory of pressure producing results.

While much of that sounds like common sense, using interim and major milestones and providing a clear vision (where you are going) and mission (what you do, along with boundaries) are practiced by very few companies. Many program or project managers agree with the concept but do not really take clear steps to practice it. The data show that these concepts are real and, finally, measured for all to see. Taking the concept further and adding a few wild assumptions, one could also put a dollar value on the strategic benefits of "opportunity gains" realized because of a good DM process such as proposals won via a better approach, better-run programs, lawsuits that were avoided because of a clear link to requirements, and so on.

Take, for example, the story on Genentech described so well by Betsy Morris, *Fortune* senior writer. The story, "Genentech: The Best Place to Work Now," appeared on CNN Money.com (January 1, 2006) as part of the "Fortune 100 Best Companies to Work For" series. In the article, with the backdrop of Genentech's fantastic growth and meteoric stock price increase on Wall Street, Morris states that CEO Art Levinson's greatest concerns revolve around culture, efficiency, and finding people absolutely dedicated to the singular focus of bringing meaningful (not just "me too") drugs to market. Per the article, "He [Levinson] says, 'The thing I worry about most is managing our growth.' And protecting Genentech's mission, focus and culture. 'It's much easier to get alignment when you have fewer people.'" Here Levinson is talking about a singular vision and mission acted on consistently by every employee, balanced by the creativity and individual personality afforded to every employee.

As the airplane began its decent into Chicago, I wondered about the groggy lady next to me. Will she ever understand the significance of that 7 percent, or will she wander aimlessly as those without a good DM process do?. Yes, the ability to make good, clear decisions can help in all aspects of business, including the enactment of an innovative thought or design.

Her question lingered, though. I can prove good DM improves overall return on employee activities (and reduces stress, as we will discuss later), and I have a postulate that it helps enact innovation, but do I have proof? And what does she define as "innovation"? What do other people consider innovative?

THE HOWARD HEAD STORY

The story of Mr. Howard Head is so all-encompassing that it could be used instead of our engine analogy to explain most of the concepts in this book. We focus on just two here. First we look at the DM of Mr. Head, including his success and failures. Later in the book, when we discuss invention and patents, we look at his incredible inventive talent and process.

Howard Head embodies the classic "build a better mousetrap" argument for invention, design, and business. There are many who argue that building a better mousetrap will not result in anything; that businesses are built on strategies, marketing, finance, and planning; and that the product is only a component. Howard Head's creation of two entire industries, recreational skiing and tennis, was based first and foremost on a better product. This example is just one of hundreds that shut down most of the arguments belittling the actual inventive product development process. Certainly one needs both: a good innovative product or service to deliver and strategy, pricing, distribution, timing, and so on. One must put the product and its use (by consumers, either businesses-to-business or end users) at least on a par with the other critical aspects such as delivery, support, strategy, pricing, partnerships, and so on.

Head's story provides interesting management principles and, more specifically, some interesting DM examples. Building a better mousetrap, ski, or tennis racket by using advanced materials is one aspect. Head was not constrained in his thinking by the vernacular of the day, in that particular industry. He quickly looked "outside the box" to other industries for solutions to a problem. The willingness to look both inside and outside your particular industry is key to success and is the second I-Factor:

> **I-Factor 2: A key aspect of success is the willingness to look inside and outside your particular industry.**

One of Head's early failures occurred when he went too far outside, building his "ultimate" prototype while skipping some basic engineering practices. He was convinced his product would be great, and despite his previous engineering rigor, he skipped his usual interim testing. At great expense and risk, he brought the finished skis to some ski pros for testing, and the skis all immediately broke. This set back his reputation and bank account considerably. So while the ultimate solution lay with the advanced materials of the aerospace industry (far outside the wooden ski industry), part of the solution also lay within the skiing industry. There was a reason those old wood skis were so heavy and solid, which Head originally ignored. Head did not repeat the mistake, ultimately setting up a prototype exchange program with a single dedicated ski instructor and testing more than 40

CASE STUDY: HEAD SKIS

Howard Head's story started in 1947 when he went skiing with some friends in Vermont. The skis were made of laminated wood, a marginal improvement over the solid hickory skis used for hundreds of years. According to an article by Stuart Leuthner in *Invention and Technology* magazine (vol. 19, no. 3), entitled "A Bad Skier's Revenge," Mr. Head was embarrassed by his own poor performance and was intrigued by the aluminum and plastic materials he had used at Martin Aircraft Company building B-26s during WW II. Head subsequently put together his life savings and set to work building a better ski. After consuming his entire personal savings and building over 40 prototype models over a period of three years, he had finally perfected the aluminum, plastic, steel, and glue sandwich that would ultimately launch some 4 million people into skiing in just a few years.

Previously, skiing was a sport for the elite, for the superrich, or for brawny adventurers. Few, if any, women were involved because of the tremendous size of the wood skis and hulk required to make them cut into the snow to turn and maneuver. Head skis changed the market for skiing. The average athlete could safely enjoy a day on the slopes. According to the article, by 1954 Head's fledgling company turned its first profit of $1,200 after years of losses. A few years later, Head was selling hundreds of thousands of skis to the public and to professionals. Head innovated in both the areas of marketing and distribution. Howard Head sold his last shares in Head skis to AMF in 1969 for $16 million. Later, working with Prince sporting goods company, he repeated the story, nearly identically, with the tennis racquet.

models to failure, on the way to perfecting the recipe. Each failure of a prototype involved hundreds of interim decisions, which Head documented thoroughly. He took notes on the thickness of materials, glue temperatures, lay-up methods, costs, and so on. He was truly trying to optimize the system, to end up with an affordable, manufacturable, sellable ski that performed better than anything else.

A second aspect of Head's exemplary DM style was his willingness to adapt. Adaptation of materials and processes from the aerospace industry and adaptation of advice from the ski instructors ultimately led to his success. He also adopted a single-minded focus on success, leading us back to the importance of vision and mission.

What I find most intriguing is a very subtle concept Howard Head exemplifies. This concept is described as inventive DM. During the process of designing any object or service, there are tens or hundreds to thousands of micro interim decisions that must be made—little forks in the road that must be taken to get to the end. Sometimes we take a fork correctly and it leads to the "eureka moment" and ultimate success. Some forks are much more mundane. Some forks lead to small failures, and the process must back up and take the other path. Mr. Head undoubtedly encountered hundreds of these during the process of building all those prototypes. He had invented a novel vacuum bag laminating process for gluing all the disparate layers together. According to Leuthner's article, Head says of the process, "It was tough work, dangerous and messy as blazes . . . The smoke and smells were terrible, anybody passing by our shop at night must have thought we appeared to be more like alchemists of the Middle Ages than men of the twentieth century trying to build a modern ski." From these humble beginnings to the point Head arrived at the correct design, there were thousands of technical, financial, strategic, and market-oriented decisions. There were discussions with landlords, unpaid employees, ski instructor testers, and so on. All of these discussions and design iterations required that the decisions be weighed against the vision—to free the world of wood skis—and the mission—to build a better ski. They were made well, and very quickly at that. Head's efficiency of design and process is legendary, and his ultimate success is a testament to it.

Finally, some other traits must be mentioned to round out the story. Clearly, Howard Head was enthusiastic. He imparted this enthusiasm to others around him, at one point making just one other true believer. This true believer was Neil Robinson, the ski instructor that tirelessly helped Head test his prototypes. Head was also willing to sacrifice everything: his personal savings, his solid job as an engineer, friendships, marriages, and so on. Closely related is his embrace of risk. Head understood risk and was willing to take on high personal risk to achieve his singular end goal. Later in life, he regretted only the personal costs of this and dedicated his later years to being a better person and husband. Leuthner then brings out one of the most interesting aspects of Mr. Head with this quote: "I'm giving up on the thing world and heading into the people world. In part, my devotion to the creative side was due to my isolation from people. If anyone ever thought of me, they'd use the adjectives *prickly* and *arrogant*. The drug of creativeness is so powerful that a person can go on and on until he dies old and lonely. I have no interest in doing that."

Howard Head's story is an amazing account of perseverance, luck, timing, and skill tied together by an attitude of innovation. His thousands of discrete decisions that gave the world usable snow skis and tennis racquets were consistent and accurate, reflecting a clear vision and mission. However, his success was not due to a computerlike accumulation of accurate decisions. It was not due to a lucky, skillful assembly of relevant decisions—some accurate, some marginal. Instead, his success, and our collective enjoyment of two sports (previously unattainable to the average person), is owed to his relentless use of innovation. He innovated in materials, manufacturing methods, marketing, market research, and even distribution and finance. Late in his life he reinvented himself and applied that same willingness to his personal life. His story is an excellent view of the overall integration of innovation and DM.

Howard Hughes is a good counterexample to Mr. Head. While Howard Hughes is one of my personal heroes, because I love his barnstorming, "anything is possible" attitude toward airplane design, many accounts say he suffered from a severe lack of consistency in his decision making and subsequent follow-through. I have personally toured the "Spruce Goose," formally known as the Hercules, and marveled at its very existence, a wooden C-5 transport 25 years before its time. However, if Mr. Hughes was a little more focused in his management style, and a little more consistent in decision making, there may have been more than one Hercules, with its hundreds of innovations flying today.

Looking at our turbine engine model, Mr. Head utilized almost every aspect of intake, compress, combust, and thrust. He had his bypass ratios and fuel mixtures correct, using a low bypass model and focusing on a few clear goals and applying his scarce resources to success and nothing else. Attuned to the feedback pressure, Mr. Head altered his path with some late-in-life adjustments to his product. One common method for capturing innovations and decisions is the use of patents. Patents and, more broadly, intellectual property help industry identify, develop, protect, trade, and exploit the decisions around innovation. In the next chapter, we explore that further.

4

INTELLECTUAL PROPERTY

"To invent you need a good imagination and a pile of junk."
—***Thomas Alva Edison***, *American inventor named on 1,093 patents, 1847–1931*

It is challenging to decipher the difference between the point where innovation starts, and ultimately produces a new product or service, and the point where a new product or service results in innovation. In other words, some innovations are linear. They begin at the intake stage as research and development and often result in a patent or trade secret, which is further refined and developed into a marketable product. Other innovations occur spontaneously, during the compress and combust phases, and create an idea or product. These innovations occur at the correct price, quality, and delivery point, capturing a segment of the market previously unattainable. Other innovations are a hybrid of linear and spontaneous elements. Without protecting either type of innovation, a competitor can easily steal the edge in the marketplace that belongs to your organization.

Mr. Head's success story as related in the previous chapter involves linear, spontaneous, and hybrid innovations. He used the protection of the patent system as well as other methods and strategies to make his business grow while protecting his products and ideas. Since the subject of intellectual property is too broad to cover completely in one chapter, this chapter focuses on the strategic nature of intellectual property and how it relates to the Innovation Lifecycle.

In the engine model, intellectual property can typically be considered the fuel, and in some pure licensing processes, it is the thrust. Intellectual property (IP) includes patents, trade secrets, copyrights, trademarks, data rights, and rights to some corporate information. It is the defining unit of what most of the world refers to as innovation. It is significant to every organization's growth and to the development of its product and personnel.

The total value of the patents and trademarks in the U.S. economy alone can be estimated to be worth trillions of dollars. According to the U.S. Patent and Trademark Office (USPTO), the interest in and use of patents are growing exponentially. The USPTO Web site lists the total number of patents issued in 1984 at 131,755. Just 20 years later, the 2004 total shows a 119 percent increase, with a total of 288,583. The current backlog of pending applications is many hundreds of thousands and growing, in part because of the business models provided by the protection and legal precedent afforded by the patent system. In an interesting study that compares and contrasts the "economic freedom" of 123 countries, the Fraser Institute shows that from 1980 to 2000, the least "economically free" countries only grew their GDP by .4 percent (.004) per year, while the most "economically free" grew by 3.4 percent per year. This is proof that economic and political freedoms, supported by enabling (and stable) legal policies and procedures, build wealth. Patent law is one of those policies.

IP can take many forms at many points within an organization and easily fits into our engine analogy. Most often, patents start with core technology ideas early in the idea gestation process (the intake section of the engine) and provide energy to the organization as they pass through every stage of the engine. Unlike products and services, patents (copyrights and, to a lesser extent, trademarks) can provide returns at every phase of the turbine process. Returns generally result from licensing revenue but can take on many other forms of value, including strategic partnerships, distribution agreements, stock and debt service, and so on.

THE PATENT VALUECYCLE™

Patents have a life cycle similar to our engine analogy but longer in duration than the life cycle of a typical product going through the engine phases. The patent life cycle starts at inception and typically lasts for 20 years. In some instances, that life cycle can be extended beyond 20 years with various patent extensions and technology development strategies, including continuation in part and other methods. By comparison, the typical consumer product has a life cycle of between six months and three years, beginning to end. Just that comparison alone shows the attractiveness of the patent and intellectual property component in an organization, the ultimate fulfillment of the "build it once—sell it 1,000 times" scenario. In short, it is a way of extending the life of a product through competitive protection. Extracting value along that entire time frame is both an art and science. The deals can range from a simple one-time sale of rights to a

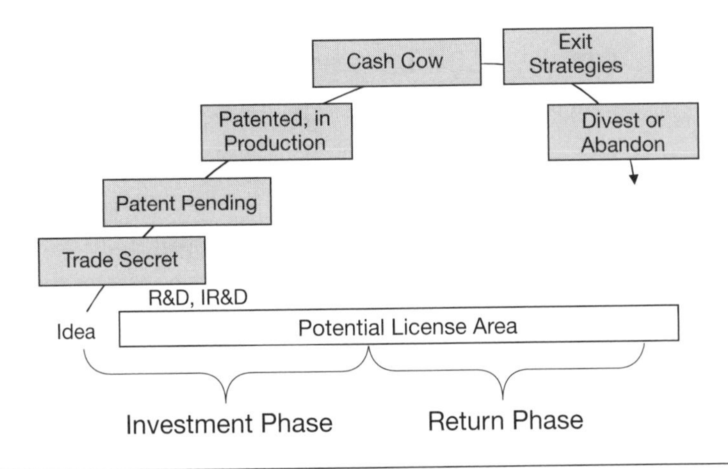

Figure 4-1 The Intellectual Property Valuecycle.

multiyear multicompany cross-license revenue share. Figure 4-1 shows a typical life cycle for a patent, also called the Intellectual Property Valuecycle™.

THE NICHIA BLUE LASER STORY

The laser is used as aid in understanding the importance of the Patent Valuecycle. Lasers power many things in our lives, from CD-ROMs in our computers and music players to eye surgery machinery to robotic vision systems for assembling toys and cars to measurement equipment. The red laser has gone through the life cycle shown in Figure 4-1 several times since its inception. Each time sales appear to be waning, new applications are developed, new manufacturing methods (which bring down the cost, increase the quality, or decrease the power consumption) are developed, and the cycle starts anew. The next breakthrough in lasers came a few years ago in the form of the blue laser.

Nichia and the blue laser can be used to explain the Patent Valuecycle. The valuecycle is a way of obtaining multiple sources of income from a single patent or technology trade secret. While I do not know the specifics of the business deals that occurred after Nakamura discovered the blue laser, it is an excellent platform from which to explain the type of license agreements that could occur using hypothetical examples. One current product on the market that uses a blue laser is the Toshiba blue-laser-driven high-definition DVD player called "BlueRay." It is one of the first to play high-definition DVDs that hold 45 gigabytes, in comparison to a regular DVD, which holds 8.5 gigabytes. The additional data equate to a higher-resolution viewing experience as well as added features.

CASE STUDY: BLUE LASERS—A STUDY IN IP

Blue lasers have nearly 10 times the data writing and reading capacity of their red brothers, a quirk of the basic physics behind light wavelength. Their "discovery" is age-old news, but the ability to manufacture them at a reasonable cost and reliability took years of research and trial and error. Most large name-brand electronics firms gave up after spending millions of dollars. One relatively small Japanese research and electronics firm called Nichia kept trying, risking nearly everything to keep funding a lone scientist on the staff who was beating his head against the problem month after month. According to *J@pan Inc* magazine, July 2001 issue, in an article entitled "Blue about Japan" by Chiaki Kitada, this little firm and, more specifically, Mr. Shuji Nakamara finally did what the giants could not and cracked the formula. The blue laser promises to transform computing technology and measurement accuracy and to realize other breakthroughs. It offers a massive increase in data storage on the same form factor as CD or DVD discs.

I must state a disclaimer now. I am not, nor ever was in any capacity, a lawyer, patent agent, patent attorney, tax advisor, accountant or other certified business and financial and legal advisor, professional engineer, or medical professional. The reader cannot infer any legal, financial, or tax advice from the stories, discussion, and examples found in this book. The reader must consult with a registered attorney or financial advisor at all times when making decisions regarding those subjects, and likewise, the reader must consult with other registered professionals when encountering those specific subjects represented by the appropriate professional. The information in this book, and in particular this chapter, is designed to provoke thought and discussion, to be applied in concert with registered professionals such as a patent attorney or tax accountant.

CARROT AND STICK LICENSING

There are two broad types of licensing: carrot and stick. I generally prefer carrot and have used some advanced strategies to succeed in that arena. However, I have been forced to use stick licensing at times as well. The difference is simple: Carrot licensing is a constructive positive exchange between two mutually interested parties (however, make no mistake, the dealings can get quite complex and crafty). Stick licensing generally takes the form of the owning party chasing the cheating party (real or perceived) over an infringement of intellectual property, copyright, or trademark. The stakes are higher in stick licensing, where treble damages can apply. The November 28, 2005, issue of *Forbes* contains an article by David Whelan entitled "Cellular Scion" on the company Qualcomm and its Code Division Multiple Access (CDMA) technology that powers most of our cell phones. Qualcomm has earned a multibillion-dollar return for years by licensing this core technology, using sophisticated patent and intellectual property strategies to extract value from its inventions in the communication industry. According to *Forbes*, CDMA licenses contribute over $2 billion a year to the $5.7 billion in annual sales. Clearly, the intellectual property strategy at Qualcomm is more than a hobby or bonus spending money. It is a key part of the corporate structure, where, per the *Forbes* article, "some 90% of this fee flow goes into earnings before taxes." Both carrot and stick are further divided into exclusive and nonexclusive deals, whereby a licensing party can choose (if offered) to buy all of the rights to use the invention—exclusive—or buy rights to use it along with the possibility that other suitors will also purchase those same rights—nonexclusive.

IBM puts such a stake here that it was granted over 2,800 patents in the year 2000, more than any other single entity. This brought its total portfolio to over 19,000 patents. In the same year, its licensing return was over $1.5 billion, and several more recent reports put the newer total at over $2 billion. For a company to get to the level of Qualcomm or other licensing giants like IBM, Microsoft, and the like, three areas must be examined: the current corporate IP portfolio, the corporate IP strategy, and the IP strategy of competitors. The corporation will likely have some old patents and licensing agreements as well as pending patents or ideas not yet fully evaluated. The IP focus of competitors must be reviewed to discern what they are protecting as a core competency and what they are outsourcing. This amounts to taking a holistic look at the company, industry, competitors, and one's own IP to enact strategically weighted decisions.

USE A TOP-DOWN ANALYSIS FIRST TO DEVELOP THE PATENT VALUECYCLE STRATEGIES

Performing a top-down analysis first, as shown in Figure 4-2, is often helpful. Various considerations must be included, compared, and contrasted against the corporate IP portfolio. These include the following:

- Corporate vision and corporate identity goals
- Division-specific vision and identity goals (if different from corporate)
- Public relations goals
- Wall Street goals and other financial considerations
- Outside market forces

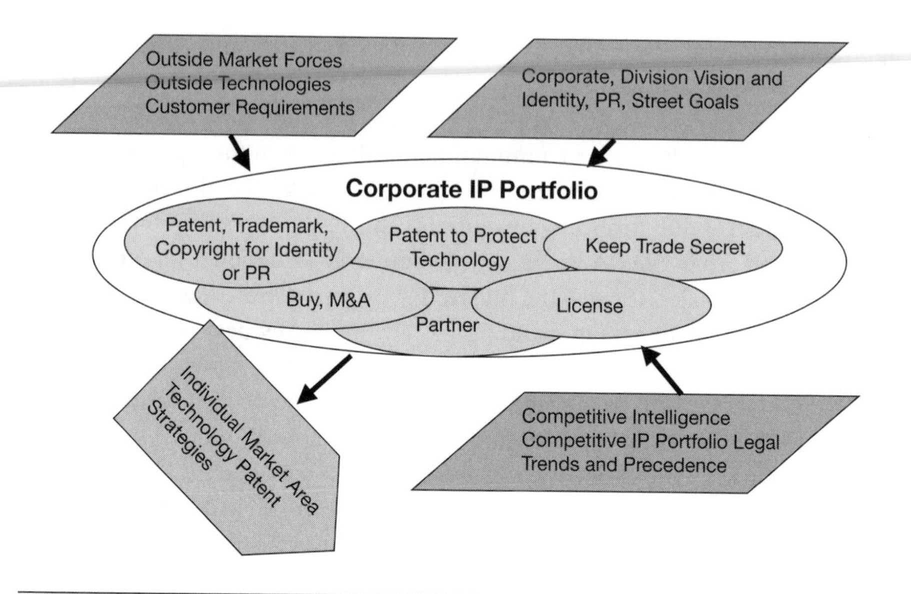

Figure 4-2 The IP portfolio analysis process. Preparing to develop individual technology area IP strategies.

- Outside technologies, industry technology trends, and other industry trends
- Customer requirements
- Competitive intelligence
- Competitive IP portfolio intelligence
- Legal trends and precedence
- Original investment
- Brand identity

Once these are considered and melded, they can begin to help form specific technology area strategies and strategies for enhancing overall corporate identity in an area. In a crazy hypothetical example, let's say that the aircraft company Boeing wanted to remake itself as a coffee shop. Boeing would then begin to develop technologies in the roasting of coffee beans and the vacuum-sealed transport of the beans and begin to divest of its aerospace technologies and IP. In this absurd example, we see how a firm can actually shape PR, whereby Boeing would then start to create news releases regarding its incredible space-age, patent-pending coffee-roasting process and so on. Patents have tremendous value even in the patent-pending stage, particularly in the area of public relations and identity strategy.

Intellectual property falls into two broad portfolios or camps: existing and future. While there is no guarantee of a patent once submitted, a firm's pending

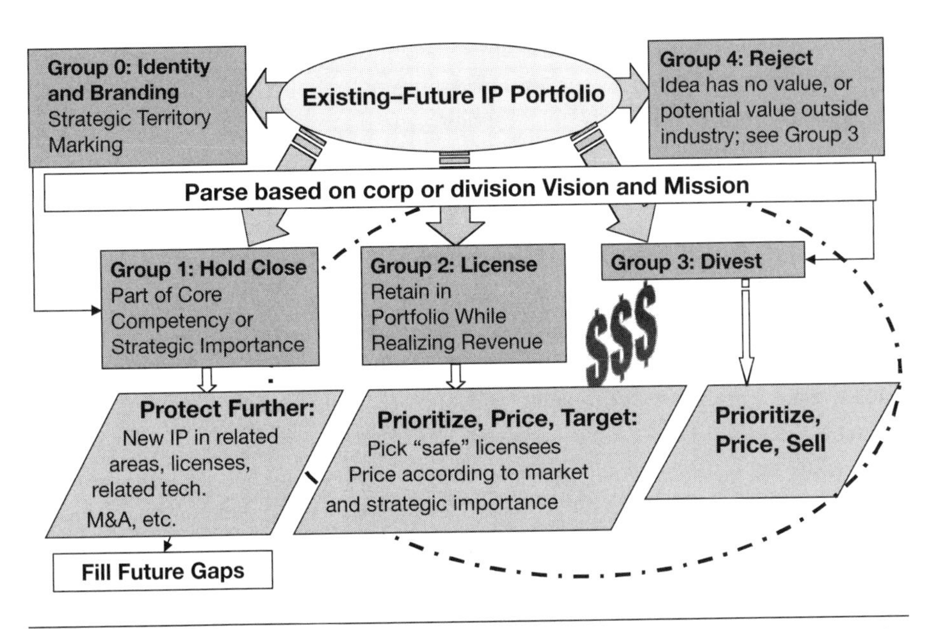

Figure 4-3 Mining the corporate IP portfolio. A straw man for mining IP value and enhancing strategic position.

applications can be broadly classified as a future pool of IP, which includes the pending applications, trade secrets, copyrights, and trademarks. This could include partnerships and the IP contained within those partnerships as well. Both existing and future pools have value.

MINING INTELLECTUAL PROPERTY VALUE AND ENHANCING STRATEGIC POSITION

The general process for drilling down IP into specific strategies can be done by technology area or by patent type and its relationship to the corporate visions and goals. Some patents are kept, some patents are enhanced, and some are divested to recoup some of the initial development costs. In Figure 4-3 we see a process for parsing and exploiting both groups.

Let's go back to the blue laser example. What if after the discovery of the cost-effective, reliable blue laser, Nichia decided to use carrot licensing to recoup some of its investment? This decision in itself is quite complex and requires a thorough understanding of Nichia's vision and mission, financial goals and needs, and so forth. It also requires that the company peer into the future, being careful not to create its own competition. Say that Nichia competes with Sony only on the fringe and has supplied Sony in the past with core and ancillary technologies. Nichia then projects out into Sony's success and failure potential futures. If Sony succeeds wildly, with high-definition, super-capacity, quad digital blue laser 3-D goggles, for example, then Nichia would receive $100 in revenue over 10 years, equating to 25 cents per goggle sold. Furthermore, these wild, futuristic goggles are not at all in Nichia's current or projected future thrust stream. The $100 is pure profit, and there is no cannibalism. Then perhaps Sony would be a suitable licensee. However, the analysis must continue and include all possibilities and dependencies. In a real-life conclusion to the blue laser story, Toshiba introduced a high-definition DVD player with five times the resolution of a typical player for the HDTV market. It boasts more detail and sharper, deeper colors for about $500, as priced in May of 2006.

Failure projections are critical for evaluating IP. Any license scenario and ultimate business decision must also include thorough failure mode analysis. Suppose in the Nichia example that Sony's futuristic 3-D goggles were a complete flop and Nichia received only the minimum contract allowance of $10. License deals often include minimums to prevent the licensor from simply locking up a technology with no intention of bringing it to market. If the license were an exclusive license, whereby only Sony was permitted to use the blue laser in a certain field, the $10 would come at the opportunity cost of other revenue. The other revenue could have been generated by licenses to Sony or other firms for additional applications. If it was a nonexclusive license, there is still the opportunity cost lost for that specific market segment. However, failure analysis must continue. A serious concern regarding the technology exposure occurs as the first products come to market. Even if Sony's wild 3-D goggles were a commercial flop, the application of the blue laser and integration into the other electronics is now public knowledge, ripe for reverse engineering by a competitor firm. Both Sony and Nichia will have "exposed" themselves in the hypercompetitive electronics industry, where copycats are just months behind the real McCoy. Now Nichia is stuck with only the minimum license fee of $10 and the chore of using a stick to chase the copycats.

REPUTATION AS A BUSINESS CONSIDERATION IN PATENT LICENSING

Patent licensing is all about understanding the risk–return universe. Naiveté here will ruin an organization's future. Part of the risk equation, beyond basic return, is a consideration of the reputation for execution of the potential licensee. The licensor must understand the licensee's reputation in that particular industry and even in that particular phase of execution. The downside risk of licensing technology willy-nilly to any potential suitor is large because failed product introductions also reflect on the core technology. Word can get around, true or not, that the core technology was at fault, and not the delivery vehicle and licensor firm. These type of factors must be weighed in an options pricing scenario to pick the overall best expected net present value out of a range of potential opportunities. Indeed, in the rather whimsical example above, we would rate Sony very high for reputation of excellence in execution and delivery. There would be some assurance that even if the basic idea of the 3-D goggles was a marketing bust, Nichia's reputation and future opportunities would not be at risk because of a faulty product or extremely late delivery. The risk would be more in the area of lost revenue and copycats.

MULTIPLE PHASES EQUALS MULTIPLE SOURCES OF INCOME

Patents are most commonly thought to apply to manufactured objects. That object might be a better mousetrap, a nine-slice toaster, a satellite radio, a camera in a cell phone, and so on. Patents protect an object a manufacturer sells from being copied or a core technology that another company will bring to fruition while paying the inventor a nickel for each copy. While these ideas about patents are valid, the typical patent is changing rapidly. Recent statistics coming from the patent office reveal the change in the quantity and type of patents being filed. Our world and the inventions that power it are changing from a mechanical to electromechanical to electrical to biological. There are now over one-half million patents in queue, many in the areas of drugs, biology, chemistry, and semiconductors and other computing electronics. Several years ago the backlog was much less and the type and usage were very different.

Table 4-1 shows the top 10 patent files for the years 1984 and 2004. This information is taken from the USPTO Web site (www.uspto.gov) and mimics similar data in other European and Asian patent databases.

THE CHANGING NATURE OF THE INTELLECTUAL PROPERTY WORLD CREATES OPPORTUNITY

The opportunity of these changes is tremendous. Our world is shifting from the physical to the biological and electronic, *and the truly innovative firms are also shifting their distribution strategies, their R&D dollars, their partnerships, and their business methods to match and win in the new economy.* While there are many variables that affect the license ability and enforceability of a patent, such as the

Table 4-1 USPTO Patent Data by Class

Rank	1984 Category	Total Issued	2004 Category	Total Issued
1	Stock material and misc. articles	2,358	Active solid-state devices	9,636
2	Drug, bioaffecting, body-treating compositions	1,894	Semiconductor device manufacturing	6,950
3	Measuring and testing	1,817	Stock material or misc.	4,840
4	Coating processes	1,526	Chemistry, molecular biology and microbiology	4,623
5	Internal combustion engines	1,520	Drug, bioaffecting, body-treating compositions	4,044
6	Plastic, nonmetallic article shaping	1,471	Drug, bioaffecting	3,196
7	Metal working	1,375	Telecommunications	3,190
8	Synthetic resins and natural rubbers	1,327	Computer graphics processing and selective visual display systems	3,168
9	Active solid-state devices	1,234	Optical systems and elements	3,011
10	Adhesive, bonding, and misc. chemical manufacture	1,186	Electrical computer and digital processing multicomputer data transfer	2,956

quality of the original invention, the quality of the writing, and so on, there is opportunity at every phase of business from the point of the patent afterward.

A core technology such as the blue laser can be licensed at every phase of the engine:

- Intake phase: Core technology permission only

In the intake phase, the basic patent is licensed to another manufacturer, with the trade secret know-how sold separately (like consulting hours).

- Compress phase: Core technology plus know-how

In the compress phase, the core technology plus the actual line machinery and process and know-how are licensed to a partner and/or a competitor to generate separate income vehicles.

- Combust and thrust phases: Combinations of manufacturing and delivery under license of core technology

In the combust phase, the actual laser diode chips manufactured are under license and sold around the world by multiple licensors. Some customers buy turnkey packages of intellectual property, equipment, and process knowledge. Other customers buy core technology only. Still others buy the unpackaged laser diode chips and downstream process the chips into blue lasers and so on.

- Thrust phase: Combinations of all of the earlier phases in secondary applications, begin to develop cross-licenses and new IP

In the thrust phase, netted and tiered agreements occur. This is where the core technology is used as is and is also used in secondary and ancillary applications. Often the licensor continues development of the core technology by developing IP of its own. If the agreements are written well, both companies benefit. Cross-licenses can be produced in the areas where the technology begins to morph and grow into new IP.

A note on cross-licenses: Not all is optimistic in the field of licensing. There are circumstances where the licensee firm is approaching a potential licensor firm, either with a carrot or stick, and finds itself being presented with a cross-license or countersuit. The old adage "there is nothing new under the sun" applies here, where the licensor firm claims that the licensee has infringed on one of its patents in the development of the core technology. These situations often get unpleasant very quickly and are only untangled by sophisticated legal agreements and when calmer heads prevail. No one should approach licensing of intellectual property without thorough due diligence. This area can get quite sophisticated; some companies prioritize their portfolio of patents by looking at the patents with the most citations by other patents. The logic here is that more citations mean more times that patent has been referenced by other firms attempting to patent something similar. Those citations are akin to a list of targets for lawsuits.

- Complete turbine process: Parlay core technology intellectual property and trade secret know-how into an empire

In reality, all of these types of agreements and IP leverage could occur simultaneously. In the best circumstances, the firm can time its release over a period of years for a constant stream of income as the technology and industry mature.

Some firms have been able to do this and achieve consistent double-digit top- and bottom-line growth based in part on IP and a good understanding of the maturity phases of its core technologies and the target industries of its application, hence overlapping patent value cycles.

It would be a disservice to provide a single glimpse into this exciting world of intellectual property innovation. I therefore encourage the reader to read the stories behind IP in any industry. Some great starting stories to research include IBM, MIT, Polaroid, Ford, Delphi, State Street Bank, Dow Chemical, Seagate Technologies, Microsoft, and Advanced Micro Devices (AMD).

Are process patents the new empire builders? The jury is still out on process patents, also known as business process patents. This type of patent was introduced by the landmark State Street Bank case in the 1980s. The USPTO is still figuring out how best to review them, how best to enforce them, and how to decipher what is truly novel and useful and what is just a lot of money spent on a patent for a method that is already well known in the industry. While process patents evolve in the coming years, firms should seriously consider this new area of burgeoning IP. One innovation in the area of business processes can often be applied to entire industries, where the potential for license revenue is tremendous. In fact, an innovation in business processes for one industry can often be applied across multiple industries if the patent is written with the appropriate hooks, increasing the revenue potential manyfold.

I hit the computer Save button after roughing out my chapter on intellectual property innovation and thought, Aha! I got her on this one. Intellectual property, the wonderful stability of the U.S. economy, and our somewhat stable and predictable legal system constitute the engine that hums in the U.S. $11.4 trillion economy. I will e-mail her this chapter and tell her that it is the "answer" to the question, why we exist, who created us, and why innovation is important and how it is intertwined with engineering and decision management.

Three days later I got her reply. "John, you are correct. Intellectual property is a key aspect of the engine in the United States, European, and Asian economies, but it is a tool, a building block, a lever, and a spear; it is not "innovation." It is the result and application of innovation. A patent is the protector and extractor of innovation's value stream, but it is not innovation in and of itself! I do think the existence of patent law helps motivate people to innovate to some minor degree, but it is not the primary driver.

"Keep digging.

"Warm regards,

"Hildie"

5

PROCESS INTEGRATION

"A system is a network of interdependent components that work together to try to accomplish the aim of the system. A system must have an aim. Without the aim, there is no system."

—***W. Edwards Deming****, American statistician, 1900–1993*

There is a link between decisions, inventions, and innovation. In fact, this link was so simple I almost missed it. A good decision management process allows the overall process to progress to the next step, phase, or milestone, at the right time and with the right mixture of elements. It primes the next stage so that it is set for success. Decision management allows any innovation that occurred in step B to flourish, take root, and become part of the baseline prior to going to step C. This is the Innovation Lifecycle, a blending of business and engineering by phase throughout the entire organization. It also leads us to the next I-Factor:

I-Factor 3: Innovation must be constant and everywhere in your process.

Perhaps the decision-making process is not always linear. However, it is impossible for innovation of any sort, process or product, to be a benefit to the organization if it is compromised by ineffective decisions or, worse and more common, by a lack of integrated decisions. Companies, families, and people in general are consistently hamstrung by the inability to make decisions. Even the most basic decisions often paralyze people and organizations. Worse yet, most decisions are nonintegrated.

Our brief e-mail exchange was the most direct and coherent we had had to date. I then e-mailed her, hoping to get more, and asked to meet at a coffee shop at Washington D.C.'s Reagan National Airport in the coming weeks. She agreed to meet but only for a brief talk. The talk was briefer than I needed, as both our planes were late getting in and she had a connection to catch.

We said good-bye and exchanged business cards. "Do you go to D.C. often?" she asked. "Yes," I answered. She was in health care, so she was also in D.C., often to schmooze with the government elite. Perhaps we would meet again, and it was left at that.

THE TUCKER STORY: A STUDY IN DECISION INTEGRATION AND VISIONARY TEAMWORK

Preston Tucker, like Howard Head, had what it takes to bring a totally revolutionary product to market. In the late 1940s, Mr. Tucker, along with some help from a few friends, set out to build and sell a revolutionary new car to the world. His goal was to build a car with better handling, better fuel efficiency and power, and better safety and performance than any other on the road. His product, and the forces acting against him, were much more complex than the single sport product in an unfounded industry that Mr. Head faced. Tucker was competing against a giant with a slingshot and only one rock. U.S. auto industry giants such as Ford, Dodge, and Chevrolet were well financed, were well established, and were invested in retaining their market position. Although Tucker ultimately failed by world business standards, he exemplifies a true innovator and good decision manager. He used a high bypass turbine model, focused wholly on one vision—to change the automotive industry for the better—and on one mission—to bring to market a revolutionary car.

Ultimately, Tucker only made 51 cars. The cars had a perimeter frame for safety, a rear engine to put weight on the drive wheels, aircraft technology to produce better fuel economy, better brakes, and countless other advanced features. Today his cars are among the most valuable on earth (some are valued at over a half million dollars). Considering the technical, political, financial, and industrial forces against Tucker and his crew, those 51 cars represent a truly amazing feat. What did he do to launch a car that was 30 to 50 years ahead of its time—a car with safety, speed, power, economy, handling, and comfort all beyond what people expected at the time? What traits did he have, what methods did he use, who was his lead

CASE STUDY: INTEGRATED DECISIONS

Suppose there are two independent decisions that need to be integrated. The decision to get milk for the baby must be integrated with the decision to let your teenage daughter borrow the only car. If she is out with the car, you cannot get to the store to buy the milk, and the baby cries. A is dependent on B, and B is independent of A, so only an integrated, planned decision method that accommodates the dependent portion will resolve the conflict. Newton referred to this relationship as basic cause and effect. The optimum decision is not one that is based simply on a solution. There are several solutions to the preceding dilemma: Delay the daughter and have her watch the baby while you run to the store for the milk, ask a neighbor, have the daughter buy the milk on her way out, and so on. Solutions are generally easy once you understand the problem. This is the very nature of cancer research, where people dedicate entire careers to discovering the cellular receptors that allow a cancer to grow. Once those are identified, the method to interrupt them—stopping growth—can be developed. However, the researcher does not start trying to kill the cancer; he or she starts with trying to grow it. *An integrated decision method that considers dependencies and is weighted or influenced by the vision produces more efficient results than a decision that only takes into account a point solution.* Unfortunately, both human nature and the need to make decisions quickly often contradict or preclude deeper, more integrated thought and analysis.

designer, his confidant, his strategist, his accountant, his general manager? What drove the small, underfunded team to take those personal and professional risks when everyone around them said it could not be done and, worse, *should* not be done.

At one point in Tucker's development cycle, he had to leave the team to go to Washington to fight some political and legal fires ignited by his enemies in the established auto industry. While he was away, his team was racing toward a deadline and came across a major problem with the "steerable" headlight subsystem. The team had a simple choice: Wait for Tucker to return and miss the deadline or make decisions in his stead and forge ahead. They were able to make decisions consistent with his desires because they knew his vision and mission, and they knew he trusted their judgment. They knew his vision was to have all of these advanced features, but they also knew that the mission to produce the car would be greatly compromised if they did not go with the simpler system for the time being and keep things moving. The compromise was the only smart answer and was made correctly. Tucker returned and the car was finished on time.

The team prevailed because of Tucker's management style and a common goal. A similar example is found in Tucker's exterior design and approval cycle, which took only seven days from design to final approval. By comparison, even the best auto firms today, with modern computer modeling, take months to approve final designs. The Tucker design achieved a coefficient of drag of only 0.27, exceptional even by today's standards.

THE TUCKER TRAITS: MANAGEMENT TRAITS FOR SUCCESS

Where would your company be if it had the following Tucker traits?

- Dogged, unflagging determination
- Clear vision with a semiflexible mission
- A rare gift of delegation with trust
- Loving, supportive family life
- Charismatic, honest character
- Innovative business and technical thinking

LESSONS LEARNED FROM TUCKER

The Tucker story is one of the best all-around examples of an integrated innovative management "system." An integrated management system has the following four attributes:

- *Clarity.* The system, goals, and process are clearly visible to all employees.
- *Efficiency.* The system must have adequate but not excessive checks and balances (all have value added).
- *Authority.* Each decision is only made once (at one level of management) and only checked once (by the next-highest level of management).
- *Repeatability.* The system must work the same at every level in the organization—whether the decision is generated from the bottom up or top down. This is the Holy Grail. If an audit were done on any decision in your organization (from buying office supplies to a major strategic product introduction), would reasonable people from the top to the bottom come close to the same conclusion?

All four of the Tucker attributes unfold in the I-Factors and the success factors linked to the principles of Intelligent Innovation.

CHAPTER REVIEW

In this chapter we covered the importance of integrated decisions to the innovative process. In order for innovation to move from phase to phase, accurate, timely, and relevant decisions are needed constantly. This is shown in I-Factor 3: Innovation must be constant and everywhere in your process.

Decisions are directly linked to management traits. Both Mr. Head and Mr. Tucker had management traits that began with a clear vision and goals and ended with the integrated nature in which they made decisions.

DISCUSSION QUESTIONS

Ask yourself the following questions. Where the answer is yes, record it and think of ways to capitalize on or improve that area. Where the answer is no, take three real steps to improve in the next month.

Are you focused?

1. For major decisions your firm is faced with, are the goals of that decision clear?

Do your processes support the need for integrated timely decisions?

2. For major decisions your firm is faced with, are the systems and process that would enact that decision clearly visible and understood by all employees?
3. How many checks and balances does a minor decision require? More than two and you are in trouble. Does each check add value?
4. How many checks and balances does a major decision require? More than three and you are in trouble. Does each check add value?

Note: Defining "minor" as a dollar value such as $10,000, defining a time value such as one week, defining a personnel value such as five employees, and so on is rather subjective and must be defined by each firm and each industry. Likewise, defining "major" also requires individual boundaries set up by the firm.

Are you integrated?

5. For each decision made, how many levels were involved? A level is defined as a management layer, title, or grade and is counted to include the person or group actually doing the work.

Each industry has its own requirement here. Hyperefficient consumer-oriented industries may be very responsive, delegating decision making down to one degree of freedom. Each decision is only made once (at one level of management) and only checked once (by the next-highest level of management). In industries where life and death are at stake, such as a hospital or airline, there may be several more layers of approval required. In any case, the key question is this: Does the person reviewing this information have the same data as the one above and below. If the answer is yes, ask what value they are adding. What additional information are they acting on? What do they bring to the table, so to speak? Assuming that the person making the original decision had the same clear view of the organization's vision and goals, wouldn't he or she make the same decision as the person above, given the same information?

Conversely, if the answer is no and the next level of management is acting on new or different information, ask why. Why is one level of management privy to information that the person or group actually charged with implementing the activity or decision is lacking? Shouldn't the people responsible also be the people with authority? This is a key measure of integration and effectiveness in any organization.

Reduction of just one review cycle, or just one layer of management, improves the effectiveness and integration of an organization by the percentage of $(x - 1) / x$, where x is the total number of levels of decision.

6

NOTHING UNWORTHY: CHOOSING HOW YOU MANAGE FINITE RESOURCES

Nil Indigne = Nothing Unworthy

This family motto from James Francis (Frank) Hurley, the early-twentieth-century explorer and photographer of the famous *Endurance* expedition, best describes the important subject of managing finite resources. Success of almost any program or project requires a constant monitoring of expenditure, which is a basic discipline of project management. In the case of a particularly innovative project, pushing into new territory at light speed, the practice of resource management becomes very complex. This complexity is due to the quantity and criticality of interrelated, often undefined capacity dependencies and limitations. The dependencies and limitations are in the areas of materials, people, funding, time, ideas—anything that goes into the funnel. Even the number and personalities of people involved and the perceived and real needs within an organization become capacities to be managed.

The people part can be the most difficult to predict, quantify, and manage effectively. Most people involved in a hot research project or a new product design consider every step along the way to be the "most important thing" or "one that could lead to a patent." They are people excited about their work and the prospect it represents. While enthusiasm for any project is good, managing it to completion is not easy. Convincing team members that there are limited resources allocated to their project can be unpleasant. This is especially true for some brilliant researchers and designers, who can be stereotyped as (necessarily) myopic and short on patience for business or organizational issues. It is difficult to keep the creative thinker motivated and managed in a singular direction while maintaining a hold on the project.

Unfortunately, managing employees is just one aspect facing the manager trying to juggle multiple "innovations" and development projects. Often the manager too is a resource that is overloaded with demand from a seemingly infinite number of directions. Similarly, organizations often have one or two specialized pieces of test equipment that are used in an overcapacity state. Money is always in short supply, schedule demands from customers or the competitive market can be crushing, and so on. All of this mayhem is exacerbated by the manager's lack of a crystal ball. Since innovative ideas and projects are, by definition, new and unknown, there is no easy metric for *xyz* variables (time, funding, material) to plug into a formula when one is trying to plan or project outcomes and balance needs. This aspect is particularly frustrating to senior management, who are trying to stitch together multiple projects and balance resources at the aggregate level.

Smart expenditure of time, energy, money, power, people, and even thought is a key Intelligent Innovation principle. *In short, spend where it counts; save the rest for later*. Do nothing that is unworthy of your goals, unworthy of your ethical framework, unworthy of local and national laws, unworthy of your precious R&D dollars, unworthy of your industry's cutting-edge needs, or unworthy of your employees' time. This principle may seem to be in sharp contrast to the practice of giving employees "free time" to create and be innovative, away from the strict schedules and requirements of established programs. That practice is neither contradictory nor incompatible with a managed resources mind-set. It is only after understanding overall resource loading that a manager can have success in mainline activities. With balanced resources, a manager could carve out 5 or 10 percent of the workforce's time for creative projects and the tip of the funnel.

In a *Fortune* article entitled "Genentech: The Best Place to Work Now" by Betsy Morris, cited in a previous chapter and that appeared on CNN Money.com (January 9, 2006), Morris describes the payoff from the Genentech policy of 20 percent discretionary time. In one example, a scientist named Ferrara discovers a blood vessel mechanism called VEGF, "launched with a breakthrough made on discretionary time." This discovery ultimately led to a drug called Avastin that treats some cancers. According to Morris, "Avastin, approved in February 2004, had sales of $774 million in the first months of 2005." That discovery would probably make Mr. Hurley's list of worthy endeavors.

So *how* does an organization decide what is worthy and what is not? What is the measuring stick? More importantly, *how* does an organization apply the measuring stick? How does an organization compare apples and wrenches, both of which necessarily coexist in an organization and both of which demand generic resources like time and money? The ability or lack of ability to make these resource decisions directly affects the overall success of the organization and will either help or hinder innovation within projects and programs.

This brings us to the fourth I-Factor:

I-Factor 4: Decide what is worthy.

Innovation cannot be legislated or scheduled, and yet the SRM Survey showed a link between R&D funding cycles and success. While much has been written on the amount of R&D dollars in percentage of total sales, percentage of total development dollars, dollars per employee, and so on, the research revealed something a little hidden but more telling. For innovations to succeed, they must be nurtured the entire way. They must be allowed to mature from the early thought phase to a drawing, to the prototype, and ultimately to the customer. Those precious ideas going into the funnel, the life-giving air of the engine, must be propelled throughout the entire compress and combust cycle in order to contribute to the thrust at the end of the cycle. Only by completing the cycle do they contribute to the resources needed to get the next idea through, and only by managing the intake, compress, and combust phases do we get to the thrust phase. Resources *in* must be managed to be fewer than resources *out*, and yet they must be enough to keep the process going, a delicate balance to be sure.

The SRM Survey asked the question "how often is your R&D budget refreshed?" The resulting data were heavily skewed, where 85 percent of the respondents answered annually. Of the 15 percent that answered greater than annual or "ad hoc" (based on milestones or other events), they were 60 percent more likely to be viewed as "successful" firms in the public and market eye. The research did attempt to quantify the term "successful" for a variety of reasons. Certainly it could have measured stock performance or return on investment (ROI) or other financial measures on this minority, struck an average, and compared it to the average of the larger body. But that type of detail was not required. What is required is an appreciation of the value of perception in the marketplace. The value of perception is immediately apparent in a firm's bottom line, the type of news coverage it gets, the number of class 1 retail outlets that carry its products, and so on.

Most R&D funding cycles, such as an annual budget, are linked to the calendar, just like everything else in the organization. Sometime near the last quarter of each year (fiscal or real depending on the organization), managers and directors of the firm get busy developing budgets for the coming year. This process includes many good and necessary things, such as justifications, pruning, resource management, past performance analysis, goal and milestone reviews, and so on. The

R&D community is usually subject to similar exercises, sometimes going through even more excruciating detail to justify their expenditures. Fine, we all accept that process as a necessary and established part of business. But what about innovation? What about those ideas that are entering the funnel or are partly through the compress cycle? They are not yet up to their proper compression ratio and not yet ready to be bypassed out of the engine either. They are stuck in limbo at the end of the year—funding over for the season and not yet turned back on for the next—regardless of the milestones approaching, the criticality of the research stage, the length of the test underway, and so on. These fledgling ideas and research products are difficult to explain technically, let alone financially. They are unknown entities, with vague, if any, tangible milestones to pin on a schedule. Their resources are needed just as badly as any established program, yet they lack the tangible ammunition to go to bat for themselves. They have no ROI. Here is where the yearly scheduled R&D budget makes no sense. It is an innovation killer.

Many R&D managers will scrape and beg to continue funding on their most critical projects. They do this so that a four-month test does not get stopped in the middle just because of a calendar-oriented financial accounting issue. This scenario should not occur. R&D funding cycles should be linked to major milestones versus calendar dates. Some would even argue that all research should be "ility-based" in its inception, including capability, survivability, affordability, manufacturability, and marketability. This "ility" focus is research that is being done to propel the firm to a certain capability, new or improved, versus a specific product—a very powerful concept. In this way the funding would be linked to the development of the capability (presumably preidentified and targeted by high-up smart people as something critical for the firm to accomplish), regardless of the end applications. It is usually a capability that relates to the core value or vision of the firm.

For example, Volvo may fund years of research into head-restraint safety for cars. This ility—call it "safety-ility"—will have broad applicability across Volvo's entire line. The first results, done irrespective of manufacturability, may be more expensive to apply. These only make it into Volvo's top-of-the-line cars or into a car that is up for redesign. As the research progresses, more and more models are added as affordability research catches up. If that type of research were strictly tied to a technical, financial, or cost milestone, it would have been killed prematurely, but instead it was tied to corporate vision and goals.

Now, again, I have oversimplified and stereotyped to make the point; these same principles can and should be applied to any innovation anywhere in the Innovation Lifecycle. My own experience suggests that the most important innovations occur far into the turbine process. An innovation in manufacturing, packaging, or service delivery method can actually be *the* difference between ultimate profitability and success and failure. This occurs long after that brilliant idea was rendered on the proverbial napkin. It is much more difficult (and radically innovative) to change the entire financial model of the organization to fund all development—wherever it occurs—according to the individual customized milestones of that development, rather than according to traditional calendar-oriented milestones. If a brave firm ever tried that, I would suggest it would end up with a portfolio-type model, with innovation capable of being funded according to regularly accepted GAAP principles, along with all the other functions of the organization, with a little extra effort.

Funding is just one type of resource. It is the easiest resource to point to and work with. Personnel, machine time, computing time, partnerships, and other capacity-limited functions are even more difficult to pin down and distribute fairly. Deciding what is worthy is more difficult in this case. As with most decisions, deciding what is worthy when allocating people, time, and equipment can be complex and must take into account multiple variables. There is a method to weigh out requirements of disparate projects and put them on one sheet of paper. This method compares the needs and benefits of multiple related and unrelated projects and helps managers decide how to allocate limited resources. This method makes the total organization operate in a unified, efficient manner directed toward a singular vision and working as if the whole organization is on a mission. The method is called *Strategic Balancing*™.

Strategic Balancing was originally developed to help firms parse scarce R&D dollars and resources (smart people and test machinery) in their yearly internal R&D budgets. As we have discussed, this activity is difficult and is often unsuccessful. Sadly, the projects that get funded are often "pet rocks" of executives or simply the ones that had the best presenter at an executive meeting.

Managers need a method to consistently bring innovative products and services to market while simultaneously tending to the boring, everyday activities of the back-office business. Companies that are able to do this are (or were) revered, such as 3M, Motorola, Texas Instruments, IBM, and Lexus. They had balanced the short-term (financial and market) needs with the longer-term goals of primary research and product development. This practice is even more difficult with longer R&D cycles. *The longer R&D funding cycles, where funds were tied to the research goals or milestones instead of strict financial calendars, were, on average, more successful.* If I had to point to one single success differentiator in mid- to large-sized firms, I would pick this one. The ESRM Survey has some data that suggest this, however difficult to quantify. My personal experience leading several major R&D programs over the years and consulting with many high-tech firms also confirms the suspicion. Firms that understand this generally survive and flourish.

Small firms are excluded for a number of reasons. Their dynamic is different, they may only have one or two projects, and they are constrained by design to the urgent quadrant. This is not a good or bad thing; it is just a matter of physics, where Momentum = Mass × Velocity. Their velocity is generally greater than the large firm (although thanks to some great books on the subject and some pioneers in the field, that is changing), but their mass is so small that they do not achieve a level of momentum necessary to enjoy the luxury of working on both long- and short-term projects. If one takes an honest look at their portfolios, the data are skewed far to the left, where they are fixing last year's warranty products, working with last quarter's deliveries, working on next quarter's development, and doing research for next year's products all at once. There is lack of mid- and long-term activities in these little firms. Activities such as industry association lobbying, mid- and long-term R&D efforts and primary research, 10- and 15-year plans, and so on are few and far between. There is no luxury for that, and there is therefore no planning and balancing of it.

Mid- and large-size firms have the luxury of a full portfolio. The smaller companies, which often spend an even greater percentage of their resources on R&D, typically do a better job of capitalizing on every innovation because there are fewer projects and those projects are more directly tied to the business at hand. The Strategic Balancing method is good for all but probably more applicable to mid- to large-size firms. For that reason, another method is also presented, a streamlined version called the "rocks and jars" method. It is an easily applied method that can be used to think about how to filter projects that are competing for the same resources.

Corollary to I-Factor 4: The art and science of resource allocation demands innovative thinking and a balanced approach at every juncture.

The rocks and jars analogy has been used by many. While the high school science teacher is trying to make a point for volumetric efficiency, the pastor makes the same point about life balance and the importance of rest and family time (i.e., don't break the jar). The director of personnel makes the point to the board of directors that the engineering department is burned out from 70-hour weeks, and they need to open some requisitions for some new hires.

The lesson in the sidebar is most often linked to capacity and the breaking of the jar. However, a subtle and much more rewarding set of lessons regarding order and strategy arise from this example, as discussed in the section that follows.

ROCKS AND JARS CASE STUDY

Many pastors, grandparents, and high school science teachers have told this story. It is probably the single best management analogy of all time.

On the surface it is a simple example. A teacher walks into class with a glass jar the size of a peanut butter jar. He has a pile of fist-sized rocks on the table. He fills the jar with the rocks and holds it up, rocks brimming to the top of the jar.

"Is the jar full?" he asks the class.

Little Jimmy in the front row says, "Yes!"

"No, Jimmy," says the teacher, "the jar is not full."

Most of the class looks puzzled until the teacher pulls out a can of smaller pebbles. He fills in some of the voids caused by the big rocks with pebbles and taps the jar a few times so they settle in.

"Is the jar full?" he asks the class.

Little Sally in the back row says, "Yes!"

"No, Sally," says the teacher, "the jar is not full."

Most of the class has a blank stare now, their little minds starting to figure out the puzzle. The teacher then pulls out a can of sand. He fills in the voids caused by the rocks and pebbles and taps the jar a few times so it all settles in.

"Is the jar full?" he asks the class.

No answer now. Everyone is suspecting a trick.

The teacher then pulls out a can of water. He fills in the voids caused by the rocks, pebbles, and sand with water until it overflows on the floor.

"Now the jar is full," says the teacher.

SELF-EVALUATION AND REVIEW: BEYOND CAPACITY PLANNING

How does an organization determine the order of placement? Which rocks do you put in first? Which rocks are used first? Which are used second? What about removing a rock?

Timing and strategy, as they relate to resource allocation, are the missing element in most of the rocks and jars examples. Understanding timing and the interrelationship between time, market forces, product development, delivery, and corporate goals is the only way to understand and develop strategy.

So ask yourself: What do you want to put in first? How often do you revisit the mix? Should the mix have high bypass or low bypass (do you want more air—open space—in the jar or less)? What does open space mean to your company or industry? In the hypercompetitive electronics industry, air, representing idle capacity, may be a bad thing. But in the creative, flexible advertising agency industry, that open space may mean the ability to woo a new big corporate client because you can convince the client that your best and brightest are not burned out and are available at the client's beck and call.

What about identification: Which are your rocks? Which are your pebbles? What is your quest? This is the factor that GM got wrong when it bought the share in Fiat. It had a pebble mixed up with a rock, and the allure of Fiat's innovative products and European market access clouded GM's actual needs, which may have been satisfied by less costly technology licenses and outsourcing and partnerships.

How do you define a rock (a big important project), and how do you define a pebble or grain of sand? What makes up the core of your mission? Answering this question will help you identify the difference between rocks and fairy dust. Remember: One person's rock is another person's pebble. A thorough understanding of your own organization's business strategy is often required to tell the difference between an industry trend or fad and what your specific firm really needs to enact its strategy. This is where creative (innovative) partnerships, colicense agreements, Cooperative Research and Development Agreements (CRADAs, a quasi-governmental form of cooperative R&D in which the intellectual property and funding mechanisms are shared), and other methods of exploring a market are useful. In other words, it is easy to get caught up in herd mentality and spend precious resources on rocks your competitors have already figured out. Sometimes there is a benefit to helping your competitors by offering something they do not have in your value chain to get something they have for yours—if that is the intelligent thing to do for the season.

Finally, what order do you put the various elements in? What time do you put them in? Do you put them in at all? This leads us to strategy and tactics. The order of placement (funding of R&D projects, priority of experiments in the lab, etc.) can be the difference between being first to market in a successful target market or being last. To summarize, the question is not only which rocks go in the jar but when, and how, and in what order. The rocks can be identified, classified, ordered, and placed with justification. An Internet start-up may choose to put only two rocks in the jar: marketing and more marketing. Alternatively, an established supermarket would split its rocks more evenly—distribution, cleanliness, lighting, advertising, specials, coupons, trucking, and so on. To make those decisions, you have to know what you do and where you are going. That requires a thorough understanding of vision, mission, and strategy.

So in one example, the Internet start-up versus the supermarket, we see several basic principles:

- Capacity planning
- Vision and mission prioritization
- Strategic planning and timing
- Balance

It is the last area, balance, where innovative, creative solutions are often required to make it all work. It is the area where, when all the analysis is done, when the pencils are down, the successful person sits back late at night and ponders, looking for the common thread, the solution that ties together all the competing items, eliminates the unnecessary, and launches the necessary.

MAKING IT REAL

Some key self-assessment questions to ask are as follows:

- Can everyone in the company, at a moment's notice, tell a stranger what the rocks and pebbles are in your company?
- Do people know how their individual role contributes to the whole?
- Are projects funded by milestone or capability development?
- Do people see the link between the importance of various projects and the vision and mission and strategy of the organization—on a grand scale and on a micro scale?

If you answered no to any of these questions, there is room for improvement.

Authority and responsibility coordination is a key effectiveness measure. It relates to everything from organizational efficiency to employee morale. Are authority and responsibility directly linked in your organization? This is an easier question than it seems.

- Does the project manager have the ability to change the direction, timing, or design of the project at hand?

If the answer is yes, after she asks four other people, gets two signatures, and waits a week for resources, then authority and responsibility are not linked. If the answer is yes but she will seek approval and advice from her direct superior, then this measure is pretty good on average.

If you cannot answer all of those questions quickly and affirmatively, it is possible you need a better method for resource allocation and decision management.

CHAPTER REVIEW

The intelligent, strategic allocation of scarce resources—including R&D funding, personnel, machine time, and anything else in critical demand—is a critical factor in capitalizing on and realizing the benefit of innovation in any part of the Innovation Lifecycle. Understanding the difference between milestone- and capability-based development is a key Intelligent Innovation principle.

The next chapter examines a few of the basic building blocks to organizational resource allocation and decision making and their link to vision, mission, strategy, and tactics.

7

A PRIORITIZATION METHOD

> "The key is not to prioritize what's on your schedule, but to schedule your priorities."
>
> —***Stephen R. Covey****, best-selling author and organizational consultant*

The following method is a simple way to assess, manage, and optimize your mix of rocks and pebbles. I have found it can be done almost subconsciously as effectively as it can be done in a written formal format. The advantage of doing it in writing is, of course, the communication aspect. It is analogous to the ECU in the turbine engine. In the ECU, fuel mixture, air pressure, turbine speed, and temperature are all balanced for optimal performance.

STEP 1: LIST THE ROCKS

Rocks are major goals, not major projects. This is a difficult distinction, and many organizations cannot get past this at first. That is OK—just list what you can in your organization's vernacular, understanding it is an iterative process, and if you do it quarterly or yearly, it will evolve for the better.

Why goals or capabilities instead of projects? Because this lifts the thinking and planning up to a higher level, where corporate resources can be more effectively managed. For example, let's look at a firm that develops three-dimensional holographic displays. In this hypothetical example based on a real firm, let's say the firm has two major projects underway: a new larger display that will help it capture the disaster planning market (looking at hurricanes and tidal waves approaching land in a computer simulation) and a new (very profitable) consulting project for the Air Force. The two projects are in constant competition, and the company is in turmoil. Why? The Air Force project manager needs the same people and skill sets as the project manager working on the new display. Both are critical to the firm, but the Air Force manager enjoys considerable leverage because of the immediate cash flow and profit he represents. Key employees are constantly ripped from their seats to hop on a plane at a moment's notice to work with the Air Force engineers on their new decision support system. Conversely, the display project manager has more respect because, arguably, he is working on the future of the firm, a capability that will ultimately yield a resellable commodity. It is a high-profile product that the world can see if it makes it into the evening news.

Suppose we were to say "these are our two rocks" and all the little projects we are working on, like warranty claims, new-customer work, incremental improvements to the existing line, and so on are our pebbles. Taking that approach to balancing resources puts us in exactly the same decision-making mode we are currently in, except with a nice front-end graphical user interface (GUI) on which to work the issue. So, the GUI (our rocks and jars construct) is helpful to a point; there will likely be no breakthrough thought.

On the other hand, if we list our rocks and pebbles in terms of capabilities, or at least generic market segments, we have a much greater chance of figuring out the real problem and solving it. In this case, the Air Force project is really "consulting" and the new display is really "next-generation hardware." Both are innovative, necessary, good things. What we find now is that they don't even belong in the same jar. We have essentially found that the firm has grown into two distinct entities. They use different billing structures, they have different profit margins, they have different customers, and they require employees with different mentalities. One group travels; the other group remains local. One group works on white boards and speaks publicly; the other works on multiple computer screens and speaks to no one. Currently, much of the tension in the firm is due to the forcing of these two disparate models, groups, functions, and customer sets together as one. Likewise, the pebbles fall into the two camps, with most of the pebbles in the jar with the new product line, our "legacy" jar and rocks. Now we begin to see that there could be a single firm with two divisions that are still linked: one providing operating capital for the other and one providing backroom credibility and depth.

STEP 2: ORDER THE ROCKS BY PRIORITY

Order your rocks using a simple grade of 1 to 5, with 5 being the highest priority. Use only one-half hour or less to accomplish this task. Do not make the grading into a Ph.D. dissertation in mathematics. Everyone should have a gut-level knowledge of what is profitable and what is not. If not, some basic numbers from accounting should suffice. Everyone should have a general understanding of what is short and long term, what is visionary, and what is missionary in nature. Grade the project according to its aggregate importance to the organization.

Therein often lies the rub: vision versus mission. We have discussed this briefly and the linkage is explained more in the next chapter. For the purposes of this exercise, it is not necessary, or even desirable, to know the finer detail, but I will explain the classic distinction and the resulting conflict. It is similar to the conflict between our two project managers above at the three-dimensional holographic firm. Visionary people and their projects are long-term oriented. A vision answers the question, where are you going? Missionary people and their projects are more grounded; they are focused on today and tomorrow and make today's income on today's customers with today's product line. We need both, but to grade them on the same chart and put them in the same jar is difficult. In a later chapter a revolutionary method for managing this disparity is presented. For now, it is best to have a healthy discussion with all the relevant people and grade these projects and put them into the same jar—unless, of course, there is a clear distinction on the capabilities as above versus long- and short-term focus.

STEP 3: LIST THE PEBBLES

A bunch of pebbles weighs as much as a rock but is harder to hold in one hand. It is not advisable to list every single project, subproject, process, issue, and detail in a planning exercise like this. Just list the top five or so pebbles that are currently active in your organization.

STEP 4: ORDER THE PEBBLES

Grade the pebbles on a scale of 1 to 5, with 5 being the highest priority. Again, try to do this in a half hour on a white board with everyone in the room. Keep it simple.

STEP 5: ADD A RESOURCE WEIGHT TO EACH ROCK AND PEBBLE, UP TO 100 PERCENT

A resource weight is different from a priority. It is an estimate of the total organizational resources that the project currently requires. For projects that are just ramping up, a reasonable estimate should be made for the foreseeable future. It is *not* related to priority; it is related to reality. In one firm studied, a well-regarded Fortune 200 firm, a single R&D project with four employees had consumed 34 percent of the budget for that research area for the past two years, when there were literally hundreds of employees consuming the remaining 66 percent. That project had received a disproportionate amount of financial resources, which was not readily apparent when one looked at the consumption of personnel resources. Both must be taken into account when the resources percentage is assigned.

The total resources must not exceed 100 percent. In reality, the total resources should not exceed $100 - x$, where x is the total amount of unstructured capacity and where Unstructured capacity = White space + Free time + Reserve capacity + Rainy-day trouble capacity that the organization needs and wants to function properly.

There are many opinions on the "correct" amount of this unstructured time. Innovation gurus often advocate 15 to 20 percent, and some data suggest that this time is the most valuable contributor to innovative breakthroughs. Our data suggested that firms with 8 to 25 percent had similar results. That broad a range suggests that some other factor is more important than the actual amount of unstructured time. Since our other data points suggested that a strong vision and mission and an efficient, if not transparent, decision management process were strongly related to innovative success, we must conclude that "some" amount of unstructured time in conjunction with a way of managing it is a key to innovative success. Perhaps our data have just proven that common sense is a key to innovation. In any case, let's continue with the exercise.

Step 5 is the toughest part of the rocks and jars exercise. It is one of the toughest factors this book will require of those who want to achieve Intelligent Innovation's success. Resources are limited—time, money, personnel energy, even intellectual energy. This limitation of life even applies to very large organizations like the government or Mitsubishi Heavy Industries. This limitation requires that some projects and pursuits be abandoned. The total resource percentage cannot exceed 100 percent. Some will argue that an organization can exceed 100 percent capacity utilization for a temporary period of time or can surge using outsourced or borrowed capacity. This is true; however, for the purposes of planning and thinking out a healthy balance for long-term performance, we avoid these options during this exercise.

STEP 6: MULTIPLY PRIORITY × RESOURCE SCORES (KEEP THE DECIMAL POINT IN THE PERCENT)

Multiplying priority and resource scores provides a fully loaded composite score that helps relate a project's importance to the firm to the resources it combines.

STEP 7: ORDER THE PROJECTS ACCORDING TO THEIR COMPOSITE SCORE

This step refers to your *actual* order of projects in the firm. The composite score tells which projects are actually 1, 2, 3, and 4.

STEP 8: COMPARE THE ACTUAL PRIORITY ORDER OF PROJECTS IN THE FIRM WITH YOUR DESIRED ORDER

When you compare the actual priority order with your desired order, do they match? If not, there is a mismatch between the perceived importance (Steps 2 and 4) and the resources consumed (Step 5). This is a common mismatch in most firms and is often a hidden cause of conflict and mismanagement, to say the least.

Do *not* regrade your projects to fix the problem. Try making the more difficult decision to reallocate resources, eliminate redundant or noncrucial projects, or combine redundant projects (an unsung, very powerful tool that demands an innovative approach to project planning).

A TYPICAL ROCKS AND JARS COMPOSITE EXAMPLE

Table 7-1 shows an example of a rocks and jars assessment. It is based on five projects named A, N, P, H, and Z and shows the ranking and weighting the team did to reconcile and prioritize the resources needed.

In Step 8 we see that things are in relative harmony. The two largest resource hogs are also the two projects—rocks—that most people ranked as the highest priority. However, there is a little surprise. Most of the people in this organization thought that the pebble titled H was lower in priority than the rock titled P. They also thought that the pebble titled Z was last in priority (of the top 5) when, in fact, pebble H is. Perhaps the people in the room doing this grading now understand why the managers of rock P and pebble H are always complaining. It may also explain why H is behind schedule and ahead on budget, with a cost performance index (CPI) of 1 and a schedule performance index (SPI) of 0.5. In other words, this manager cannot even spend from the budget because there are no resources—people—being made available to him.

Table 7-1 Rocks and Jars Composite Example

Step #	Project Name	Priority	Resource Weight	Composite Score/Order	Expected Order
1 and 3	A				
	N				
	P				
	H—Pebble				
	Z—Pebble				
2 and 4	A	5			
	N	3			
	P	4			
	H	2			
	Z	5			
5	A	5	20%		
	N	3	30%		
	P	4	10%		
	H	2	10%		
	Z	5	12%		
6	A			1.0/1	
	N			0.9/2	
	P			0.4/4	
	H			0.2/5	
	Z			0.6/3	
7	A			1	1
	N			2	2
	P			4	3
	H			5	4
	Z			3	5

8

THE STRATEGIC BALANCING METHOD

"However beautiful the strategy, you should occasionally look at the results."

—***Winston Churchill****, statesman and prime minister of the United Kingdom, 1874–1965*

The decomposition of vision, mission, strategy, tactics, and operations is the core to any decision. These elements provide the construct and relevance. A lot of people are working very hard on visibility of information. Project Management Institute (PMI) literature is full of ads on this, as the ultimate goal is to manage the entire organization effectively. There are better and better ERP systems, dashboards, reports, and methods—all designed to give us raw information, *but we still lack the "reason" or "framework" with which to make decisions.* Dr. Piali De of Raytheon Integrated Defense Systems says it this way: "In order to make a good decision, to run the mission well, we must first know the organization, the tasks, the roles and responsibilities and rules that govern the roles. This boils down to knowing who does what, when, why, how well, how else and what next." To propel our innovations, or at least not stagnate them with stale, pat, slow, overanalyzed, or overapproved decisions, we must begin with the end in mind. Integrating the data and the reason/framework provides a good basis from which to make better decisions. Better decisions regarding every stage of the engine and regarding the interrelationship among the stages helps the organization achieve consistently higher performance throughout the Innovation Lifecycle.

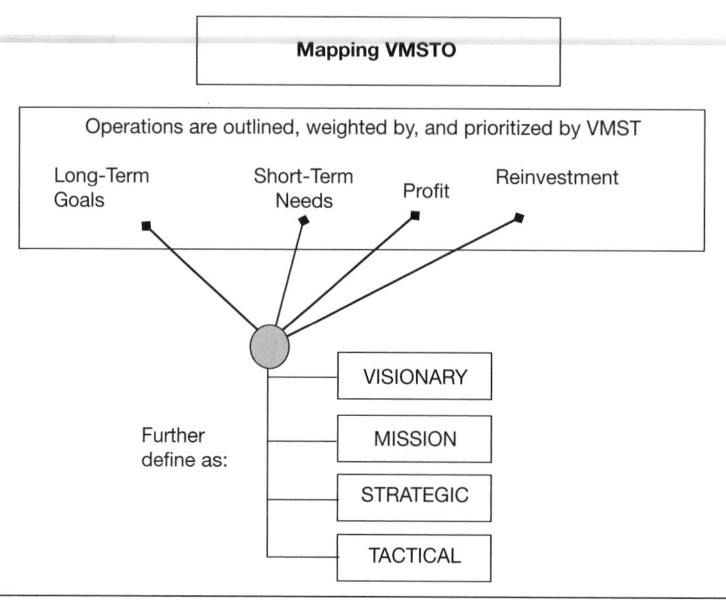

Figure 8-1 VMSTO framework.

A simple framework is shown in Figure 8-1. This type of framework helps the decision maker, resource balancer, master scheduler, program manager, and others understand what is needed and why. Some people can do this intuitively, some benefit from a graphic like this, and some need detailed tables and charts filled out with every detail. All need a gut feeling or sense of what is required to proceed, which is derived from a clear vision and mission. The full picture, including vision, mission, strategies, tactics, and operations (VMSTO), is needed just to get going. All these factors are rarely presented together. A few great leaders seem to be able to process all of these inputs simultaneously, or at least they have all the bases covered in the skills and personalities of their subordinates and systems, which, when taken in aggregate, complete the picture. Even more rare is a leader who has the ability to process all of these inputs and act on them simultaneously.

In his book *Blink: The Power of Thinking without Thinking* (Little, Brown, 2005), Malcolm Gladwell talks about this holistic gut-feeling decision ability and poses some interesting examples. Here a more structured approach for managing and extracting value is presented. It is useful for an ongoing organization with a portfolio of projects experiencing resource competition.

Strategic Balancing is a management framework that is specifically designed to help focus, balance, prioritize, and conclude enterprise-wide planning and management issues in any organization on any project. It is particularly useful in

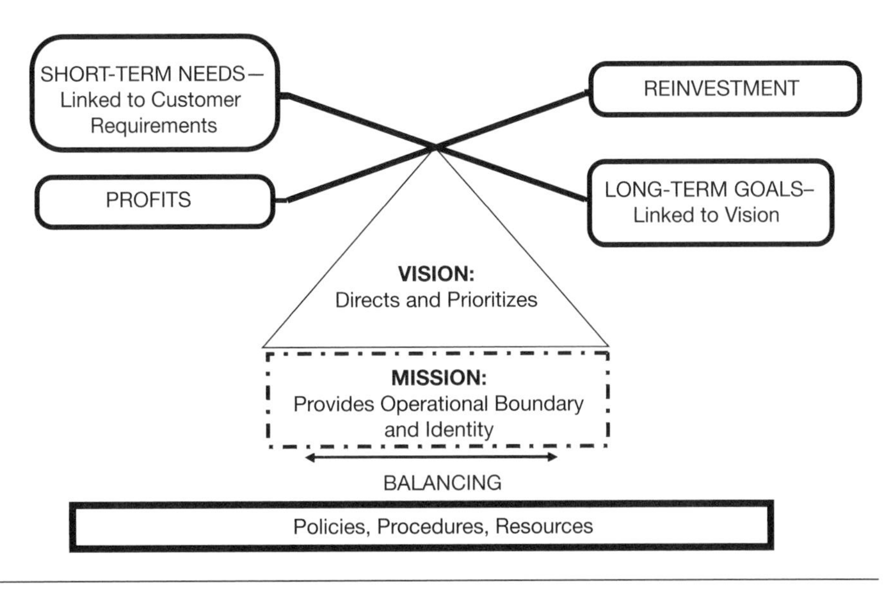

Figure 8-2 Strategic Balancing process.

large organizations with multiple technology development projects or on complex projects such as reorganizations, government services, product development, or construction. It is essential if a company is developing a new product from scratch.

Strategic Balancing helps the user develop useful, relevant vision, mission, and goal statements that can then be used to make decisions and foster innovation. These statements are used to tackle difficult questions, to negotiate the tension between short-term needs and long-term goals, and to allocate resources between revenue-producing and reinvestment-oriented activities. While we will talk about the tension between risk and innovation in a later chapter, that tension is a subset of the tension between long- and short-term goals and needs, which is a necessary part of Strategic Balancing and business. Once the forces that govern the firm and its environment are understood and a framework for the management is in place, then and only then can innovation flourish in its given pursuit.

Strategic Balancing is the only method of its kind. Based on sound business and engineering principles, it offers a truly integrated, holistic approach to complex decision making. It is easy enough for everyone to use, from the senior strategy team to the operational personnel. A simplified pictorial view is shown in Figure 8-2. The Strategic Balancing method produces an enterprise-wide integrated plan via ranking projects and subprojects by various attributes linked to a common vision and mission.

The fifth I-Factor is a result of this balancing process and was found in several success stories:

I-Factor 5: Decision-making efficiency, accuracy, and relevancy are all crucial to the economic and emotional well-being of the organization and are directly related to the dissemination of a well-crafted vision and mission.

Attributes include maturity of development, risk level, priority, link to corporate goals, strategic intent, timing, and the like. This ranking produces a composite score that allows managers to view multiple disparate projects from one consolidated point. This capability is critical to evaluating strategic plans and marketing proposals.

WHY STRATEGIC BALANCING?

Strategic Balancing is based on the concept that the outer environment (business, economic, regulatory, competitive) is always changing and each change carries with it the potential need for a decision. The inner environment is also changing, as new innovations, new employees, retiring employees, and financial pressures all demand action. The inner and outer forces and competing issues are depicted in Figure 8-3.

Strategic Balancing encompasses both inputs and outputs, providing both graphical and numerical views to the data involved in each decision. Those weights are influenced by the policies and procedures (and historical cultural elements) of the organization as shown in Figure 8-3. However, Strategic Balancing is not a manage-by-numbers schema. It is a process deeply concerned with the human aspect of management and the changing nature of business environments. It is applied through a philosophy and mind-set backed by definite quantitative metrics and qualitative measures rather than the other way around. In other words, it is easy to trace a decision back to its reasons, but the data do not dictate the decision.

THE APPLICATION

While application of a methodology like Strategic Balancing can and should be a part of everyday business, it is often associated with a major restructuring or reengineering or product development program. Strategic Balancing is appropriate for almost any organization, large and small, manufacturing and service. Personnel feeling trapped in a strictly hierarchical structure will embrace the

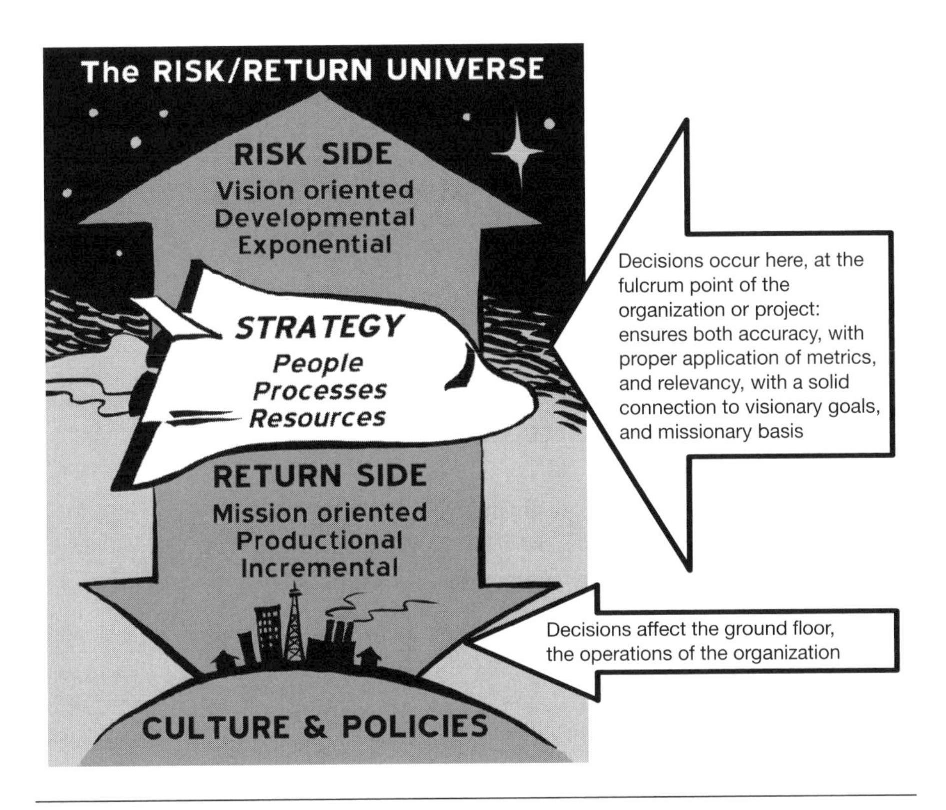

Figure 8-3 The risk–return universe.

method's ability to logically transfer authority to a point of responsibility without usurping the chain. Employees feeling frustrated by the lack of drive and focus found in flat organizations will enjoy the structure and backing afforded by the Strategic Balancing method. Nonprofits and governmental bodies have found it provides direction and motivates employees.

Strategic Balancing helps answer these typical questions:

Strategic:

How do I develop a multifaceted plan that includes our future goals and current constraints?

How do I price products and services so as to develop a strong base while not selling us short for the future?

What should my delivered products/services look like to match the needs of my market in the future?

Organizational:

How do we become a synchronized organization? How can disparate departments know what each other is responsible for? How do we balance how each department contributes to the common goal?

How do I manage limited manpower, resources, and budgets across multiple projects in multiple departments?

Technology Development:

How do I manage multiple projects in various states of completion?

What should we pursue: the need to innovate, to be radical, to produce new services/products or the need to manage risk and produce reliable profits?

How do I accurately gauge performance of multiple projects going in different directions with different goals and metrics? How do I compare apples to apples when I have nuts, berries, and rocks?

An organization's management infrastructure is composed of the various tools to develop, store, and disseminate the information required to carry out operations. It also includes the processes and procedures that employees follow while using the tools. Tool sets range in complexity from simple files to Web-enabled supply chain management systems. At the highest level, the organizational infrastructure includes philosophies and behavioral rules such as total quality management, concurrent engineering, financial goals, legal policies, and so on. At the macro level, applications such as ERP and subsets such as manufacturing resource planning (MRP), earned value management (EVM), accounting, and inventory management are used to provide the information and structure necessary to carry out operations. On a micro level, program and project management uses budgets, schedules, and risk reports to manage the daily tasks. Taken together, the macro and micro processes and all related decisions can be termed "management."

Improving the art and science of management, irrespective of innovative topics, is a constant effort for many. The 1980s brought efforts in process improvement, in business process reengineering (BPR), and in systems improvement and the beginning of the ERP stage. The 1990s brought refinements in the same areas, with a much-needed focus on integration issues stemming from the widespread failure of many early BPR and ERP attempts. Improving these systems is not easy. For every change in process, a change in culture or behavior is required. Adjusting the culture of an organization can be just as difficult and perhaps is the area most in need of innovative approaches. In his book *Leading Change* (Harvard Business

School Press, 1996), John P. Kotter suggests that corporate culture is a "residue," clinging to the organization long after attempts to change it have been completed.

Most experienced managers would confirm Mr. Kotter's residue theory. In many cases the management improvement efforts of the 1990s still failed to integrate, target, and focus the energy of the organization as a whole. The same poor decisions were just made at a faster rate. Innovations failed because they had no foundations. In some cases there was no consistent underlying process. In other cases there was no ECU. In still others there was a great combustion chamber but no turbine.

Improvements in information infrastructure may have caused a new set of suboptimizing problems and may have re-created some of the boundaries they originally intended to destroy. In an article entitled "Paradox and Project Management Culture" (*PM Network* magazine, July 1999), G. D. Githens interviews a variety of people. One project manager at a major medical device firm puts it best: "I have seen companies do terrible damage by initiating a poorly thought out project management improvement program. The worst seems to be from fragmented deployment of various strategies, tools or enablers without an integrating framework. Frustration and cynicism result." Strategic Balancing helps avoid that frustration by providing a focusing, harmonizing, optimizing method that leverages the best of the ERP and BPR trends while including all aspects of the organization's management structure and goals. It is useful to understand some of the underlying theory to decision management and capacity planning—beyond the rocks and jars example—before comprehending the entire Strategic Balancing method. Basically we know that any organization, even the federal government, has limitations, and these limitations necessitate resource allocation decisions. We also know that both the internal and external business and the political environment are always changing, also necessitating decisions and management. Together these forces require a nimble, accurate, delegated decision schema. The need for constant innovation, enhanced by global competition, just magnifies these basic forces, requiring an even better schema and application.

Strategic Balancing is based on two theorems: theorem 1 and 2. Theorem 1 is further supported by condition A and condition B. It is best to explain theorem 1 by first examining conditions A and B. Condition A states that operational and/or business environments and both internal and external forces are dynamic and always will be. Therefore, the organization must be positioned to smartly move forward through the changes. These dynamics are a basic fact of business life and form the basis for theorem 1. While condition A espouses that the most successful organizations deliberately capitalize on change and harness environmental situations where possible, this proactive response is not a requirement of the theorem. Likewise, condition A also ties entrepreneurship and innovation into the overall Strategic Balancing process. This is not a condition of the theorem, but rather a linkage. Ultimately, condition A recognizes that any pursuit, from a military campaign to management of an established consumer brand, is affected by internal and external or market forces and must be maintained and managed, no matter how stable or instable.

In condition B, the organization has limited resources in terms of intellectual energy, finances, and time, also a fact of business life. Condition B also extends to other areas such as intellectual property, product life cycle, regulatory compliance, consumer demand, societal trends, the Information Lifecycle™ demographics, and so on. There is no such thing as unlimited resources.

Combining conditions A and B produces the first theorem of Strategic Balancing.

THEOREM 1: RESOURCES AND DECISIONS

Since the global/business environment is constantly changing, decisions have a finite length of time validity and are constantly required (there are no decisionless pursuits). Because the organization is faced with finite resources, including intellectual energy, financial means, and time, and decisions affect and utilize those limited resources, then each decision must be timely, relevant, and accurate to ensure success or long-term survival.

With condition A, decisions are in constant demand because of the variability of the exterior and interior environments. With condition B, decisions are required at every level of the organization, no matter how perfect or automated the process, because of the changing nature of the external and internal environment and limited resources. Innovation in the decision process is required to obtain the ideal balanced result.

THEOREM 2: CONSISTENT DECISION MAKING

Theorem 2 is used in conjunction with theorem 1 and rounds out the balancing theory. Theorem 2 is as follows: Because decisions are in constant demand and must be made at every level of the organization, regardless of process, the organization must practice some form of successful, harmonious, and consistent decision making to succeed and ultimately to remain in existence.

Taken as a whole, Strategic Balancing suggests that innovative, consistent, and efficient decision making is goal oriented, vision and mission specific, requirements based, and quantifiable, and it does not violate process or employee integrity.

Although Strategic Balancing can be performed in an ad hoc informal manner, it can also be performed in a more structured documented way as outlined in the following.

THE METHOD

Phase 1: Preparation

In the preparation phase, a modified Integrated Definition (IDEF) methodology is used. Major processes (operations, functions, programs, and projects), inputs, outputs, sources, and destinations are mapped. An exhaustive map is not required—just one that outlines major activities and processes. If the organization has been through a business process reengineering program within the past year or so, much of that work can be reused. It is also helpful to obtain the financial, technical, or other stated goals for the organization.

An organization can use the following steps to prepare for discussions:

1. Appoint a project lead and facilitator. This person's role is ultimate decision making in the case of a "draw" and to keep the process moving. Studies have shown that groups that linger on topics too long and cannot move on to decision making and execution ultimately fail. Keep in mind that often a bad decision is better than no decision.
2. Decide on a time budget. Depending on the situation, the typical budget for this activity can be from two sessions of two hours each to four sessions of four hours each.
3. Complete the answers to the following preparatory questions prior to the first meeting:
   - Why are vision and mission concepts important to my company (or organization, family, school system, or town government)?
   - Do we see vision and mission as fluffy statements good for a plaque on the wall? Why or why not?
   - Where are we going (vision)?
   - What do we do (mission)?
   - How do we do it (strategy)?

Following are a few additional hints to get started:

1. Look up the definitions of vision and mission in a dictionary. Discuss as a group how these definitions apply to product design and brainstorming.

2. As a group, ask "what" questions: What are we doing? What are we designing? What is required? What is the product? What is the market? What are our long-term goals? What are our short-term needs? Then take the answers to the "what" questions and craft a single statement no longer than two sentences. This is your mission.
3. Ask "where" questions: Where are we going? Where is the product going? Where is the company/sponsor going?
4. Then ask the following:
   - Why are we designing this product?
   - When is this needed?
   - What are our main goals?
   - What are the product's main goals?
   - What are the sponsor's main goals?
5. Take the answers to the preceding questions and craft vision and mission statements. The statement should provide some sense of direction, energy, goals, flow, and process.

The next phase puts the details and data into the schema.

Phase 2: Detailing

Mentioned earlier in the chapter and shown in Figure 8-1, the detailing process is called VMSTO. VMSTO begins with the vision. The question "where is this organization going?" is asked and answered in several forms, as seen in the questions above. It is the vision that provides the decision-making fulcrum, a unifying object of direction for all to use. An understanding of the mission is also required. The question "what does this organization do?" is asked in several forms. Mission definition is critical; it provides a boundary and identity for the organization. In effect, by creating the mission, the organization relinquishes the idea of being all things to all people and sets out on a defined, but not definite, course. The mission is best drawn as a box with dotted lines. It will change over time but with a relatively long frequency. Once the vision and mission are accurately portrayed, the process continues, outlining the strategies and tactics within the firm.

Several reference materials are available on the development and crafting of a vision and mission statement. The reader is encouraged to use these as necessary. When crafting a vision and a mission, keep in mind that in many ways it is better to develop these statements without a lot of fanfare, use of outside consultants, or overanalysis. The best statements are usually short and clear and reflect the immediate "elevator pitch" of the senior executives.

In short, the vision statement answers the question "where are we going?" It is a long-term (a 3- to 10-year view), process-oriented statement with a goal-like slant. However, it is not a goal statement. It is a statement that provides identity-centric direction and motivation from which goals can be derived. Goal statements are also good, providing some substance to the vision statement, but they are always subordinate to the vision and build up to it. A well-crafted vision statement helps the user make marketing and design trade-off decisions, particularly in relation to timing, strategy, and resource allocation.

The mission statement, on the other hand, answers the question "what do we do?" It is not time or goal based; it is product or output based. A mission statement provides a boundary for the project; it allows the users to see "what's in and what's out" and to make timely, efficient, and more relevant decisions as a result.

The next step in this phase is to develop a strategic statement. Strategy is the "how"; it is the action-oriented part of a vision and mission. How do you get where you are going while doing what you do? It is time based, with intervals of months to years depending on the situation.

Together, the vision, mission, and strategy statements become an integral part of your design methodology, a part of your group formation, and a part of your project planning. Likewise, tactics and operations take those upper-level commands and directions and further decompose them into actionable everyday steps and to-do lists and actions.

The intake phase of the engine, also known as the funnel or concept exploration and early development phase, is necessarily a bit more loosely defined than the same statement for the same product in a later stage of development, and yet it should impart some sense of direction. There is a delicate balance here. A firm or design team should not want to limit itself to one corner of a design solution when the other corner had an innovative or elegant solution just waiting to be explored. On the other hand, one must provide enough direction with the vision statement and enough of a boundary with the mission statement so as to provide a construct for decision making and resource allocation so that the team can move forward with a reasonable amount of efficiency and decisiveness.

Phase 3: Scoring

Using the VMSTO charts developed earlier, the team progresses through a successive, multitiered rating and ranking of each element. The operations of the firm are given a numerical rating and separated by the strategy or tactic. Once VMS have been outlined, scored, and weighted, they must be used to make decisions and allocate resources through the tactical and operational levels of the organization. The level of detail required to manage is directly proportional to the level of accountability desired for management. Said another way, if we do not provide the deck plate employees with enough information to manage their work, they should not be held accountable for its execution. This is getting at the core processes and policies of the firm. The team must ask these questions: Is the mission bounded? Is the operation we are mapping in this particular step XYZ directly supported by the information required to get it done? Is the person responsible, able, and enabled to make any required decisions? There can only be either a yes or no answer to each of these questions. If no, the process or policy must be changed if the firm is to be efficient and promote Intelligent Innovation.

Using the VMS composite scores, tactical and operational activities are separated by their weighted link to the vision and mission. This process is roughly analogous to the decomposition of decision variables as described by C. W. Kirkwood in his book *Strategic Decision Making* (Wadsworth Publishing, 1997). Similar concepts are found in W. L. Winston's book *Operations Research Applications and Algorithms* (PWS Kent Publishing, 1991). The weighting and rating process produces a prioritization that sets up the balancing equation. A simplified quality function deployment (QFD) process can be used with adequate results. Good discussions of these basic procedures can be found in various texts on multi-objective decision making, operations research, business statistics, and QFD.

The balancing equation is the summary of all this weighting and grading. Every function (which now carries a value from the weight/rate process) within the organization is classified into one of the four groups: long-term goals, short-term needs, profits, and reinvestments, as shown in the framework in Figure 8-1.

Note: For nonprofits, military organizations, schools, churches, and so on, profits and reinvestment can be measured with several different metrics, including progress, capital raised, people served, and applications processed.

Phase 4: Balancing

From this point the management team is better equipped to make innovative, vision-based, mission-centric decisions in all areas, from marketing to research and development to employee benefits and facilities investment. This basis now helps foster innovation in each area. Decisions made are consistent with the vision and mission, and since priorities (weights) are known, innovative ideas can also be related to their relative priority, and an efficiency of innovative ideas develops. In other words, people stop "herding kittens" and get down to business.

Each idea, project, decision, funding allocation, and capacity plan is run through either the informal process or the formal process or both, depending on its gravity and complexity. The balancing equation helps decision making by providing a visual juxtaposition of all elements related to a decision. This holistic representation of decision variables and their interrelationship to other aspects (resources, other competing decisions, strategies, goals, etc.) is typically far beyond the level of information provided by ERP systems and typical reports. What-if scenarios are an integral part of the Strategic Balancing method. Depending on the particular industry cycle (yearly for municipalities and school districts, every two months for a high-tech start-up, etc.), the process is rerun in an abbreviated format (assuming no major swing in the organization's direction or content). In particular, the strategic section is juxtaposed against the tactical section. Strategies often become tactics, and tactics play out to conclusion as items are either accomplished or found to be in effective.

For example, a small, growing technology firm may as part of its strategy elect to spend more than its net worth on advertising and public relations in a one-time, do-or-die attempt to attract second-tier financing. For a midsized five-year-old manufacturing concern, this type of expenditure would be ludicrous, where equal amounts of operating cash are needed for operations, marketing, infrastructure, and quarterly taxes. Both strategies are valid. Both are based on the particular needs and goals of the firm at a point in time, and both must be revisited periodically. The Strategic Balancing method requires a certain amount of replanning and maintenance due in part to practicality and in part to adherence with theorem 1 and 2. The Strategic Balancing method and a facilitator help manage the entire process, along with a good spreadsheet and word processor.

The process is often best accomplished the first time using a trained facilitator. The direction and mission of the firm are first culled during the four-step process. Affinity diagrams, key word exercises, and other tools are used to focus information into distinct categories. During this initial process, the facilitator acts more like a coach than an expert or advisor. This approach helps the facilitator understand the relational and operational workings of the firm without directly influencing the outcome or hindering any team building that occurs during the meetings. It is strongly recommended that the facilitator is not a stakeholder in any core process.

CHAPTER REVIEW

Strategic Balancing provides a formal construct for organizations to define, decompose, and utilize the ingredients of a vision, mission, strategy, tactics, and operations. It is valuable during every phase of a project or business life cycle and requires people to work together toward a common understanding of the issues. It is not innovative in and of itself, but it is a supporter and enabler of innovation, particularly in large organizations if decision authority is also delegated. Chapter 9 provides a comprehensive example and review of the method.

9

THE QUALITY FOODS INCORPORATED CASE STUDY

"Quality in a product or service is not what the supplier puts in. It is what the customer gets out and is willing to pay for. A product is not quality because it is hard to make and costs a lot of money, as manufacturers typically believe. This is incompetence. Customers pay only for what is of use to them and gives them value. Nothing else constitutes quality."

—***Peter F. Drucker****, management consultant and author, 1909–2005*

This chapter provides a comprehensive example of the Strategic Balancing method based on an actual client, renamed here as the fictitious Quality Foods Incorporated. While working with Quality Foods, I was getting deep into the balanced strategic planning method when I received a call from Washington, D.C. to work with the U.S. Navy on a new type of communication system. The call was urgent, and I had to leave within eight hours or lose my spot on the development team. My duties with the U.S. Navy and its contractor involved rather mundane management duties, but the job paid a lot of bills. So I ran to my MBA class that morning, finished teaching a few minutes early, and raced to the airport, buying a tie and fresh shirt on the way.

Flying out of Logan, I take the manuscript for this book out again and remember my challenging friend's last comment: "So what about innovation?" The flight is short and I can't get her challenge out of my head. I have plenty of data, proof, and examples of the importance of innovation but have not really used them with my new client. We are focusing so much on process and policies that I have not pushed the innovation topic, nor have we directly analyzed the client's ability to capitalize on anything coming through the intake stage.

Sure enough, I see her at a coffee shop just outside the Reagan National terminal. We say a warm hello and laugh at the fate of it all. She is just starting to board and we talk quickly. "How's the manuscript?" she asks. I know she could care less about the actual manuscript, my editor's comments, and so on. I say, "I am still struggling to show the link between vision, mission, strategy, and innovation, but I think it's there somewhere."

I continue, "I look at some of the most successful firms and I keep asking myself, 'So where is the innovative spark?' I can see the value of good management, a good strategy, prioritization, a clear vision, and a mission, but I don't see how that company was innovative. I have this 'Strategic Balancing' method going and it seems to work well for my clients, but I also want to also infuse a spark into their organizations, and Strategic Balancing isn't exactly 'sparky.'"

As usual, she laughs at me and says in her movie-actor voice, "Well, darling, you're staring right at it." That's it! She walks away like we are filming the dramatic end to a 1930s film noir. All the scene needed was for her to have a cigarette in a long holder and a fur coat. The scene was already complete with the costar's dumbfounded look on his face that says, "Hey! Come back here. I have more to ask you!"

It's in there somewhere . . . it's in there somewhere . . . I keep repeating as her plane takes off. I feel like I am looking for Toto.

Strategic Balancing is less "innovative" in and of itself; it is more of an *innovation enabler*, as it helps the entity decide what is worthy and then manage that to conclusion. It helps provide a linkage among interdependent elements and then uses that linkage as a decision-making framework. Authority and responsibility are associated with long-term goals, short-term needs, profits, and reinvestments and are then weighted by the organization's vision and mission.

Strategy, tactics, and operations are controlled by the resulting plan. Proper integration of these factors with each other and with the marketplace is the critical link missing in several management practices used today. Strategic Balancing is scalable and can be applied to the areas of corporate management, engineering management, or project management. It has implications across almost every area of an organization's operations, including communications and collateral; investment in property, plant, and equipment (PP&E); retained earnings and ROI goals; infrastructure and information systems (IS) programs; sales and marketing plans; training; product/service design and manufacture/delivery; new product/service development; R&D spending; and employee growth and training.

Its most powerful application occurs not in the process itself but in daily decision making—the informal side. At this level, Strategic Balancing is accomplished without any formulas, charts, or weighting factors; it is accomplished by each person in his or her moment-to-moment decisions. I postulate that complete adoption of the Strategic Balancing methodology, carried out for at least a year, will drastically reduce the number of decision iterations while increasing the decision accuracy and relevancy. Executives and hourly workers alike can ask two simple questions when making decisions on any subject (idea, project, decision, funding allocation, capacity plan) that affects the business:

1. Does this (operation, decision, investment, or situation) match our corporate vision? Does it get us where we are going? This vision-centric question affects accuracy, efficiency, and robustness. Would an innovative approach to this make it happen faster, better, and more profitably?
2. Does this (operation, decision, investment, or situation) fit within our corporate mission? Does it get us where we are going while we are doing what we do? This mission-centric question affects relevancy, efficiency, and robustness.

If the answer to both questions is yes, then the decision point can be captured and enacted and progress can be measured against the goal. Management then begins the continual process of "balancing" it against the foundational values by using the VMSTO methodology and the mission-specific formula.

If the answers to the vision and mission questions are no, there is one of two options: (1) it is worth researching further and has value in a different form, or (2) discard it and move on, saving precious limited resources and avoiding an endless cycle of no-win decision making.

Strategic Balancing, good decisions, good policies, and great ideas are all nothing without a culture that permits innovation to flourish and take root. Indeed, all of these good practices help create the culture, but they are only a small contributor next to a deliberate effort and declaration of an innovative culture. We dive into this important topic of organizational culture in the concluding chapter.

My trip to D.C. goes well, and I receive a contract to continue with the Navy for six more months. The work is nothing top secret or very technical, but my client seems to like what I have to say and I am grateful for the opportunity. Back at Reagan National, I wander over to the watch counter and look at all the cool gear. It hits me: She was right.

FUEL: KNOWING YOUR CUSTOMER'S CUSTOMER

I was looking right at it . . . Jones Aircraft, Quality Foods, Starbucks, McDonald's, Wal-Mart . . . the innovation is in the processes, not in the actual product. At Quality Foods, innovation was in the original decision to make (and later keep) trucking part of its core competency and strategy. The decision was a sort of end around Quality Foods' competitors, who treated its trucking operation with disdain and negligence, ultimately outsourcing trucking as an annoyance and "expense."

Quality Foods saw beyond the immediate balance sheet and into the value stream. Quality Foods had realized what was the engine and what was the fuel. It understood its entire business from its customer's customer's point of view. This is intensely innovative, and it is counter accounting cultural.

Quality Foods exemplifies good process innovation, which we discuss in detail in Chapter 10.

Case Study: Quality Foods Incorporated

Quality Foods Incorporated needed to decide whether or not to outsource its trucking function. The trucking function is a major cost center and requires a large infrastructure and support organization. It is also a significant source of liability exposure when compared with warehousing functions. Trucking is also a critical element in the quality of the service, which is measured by on-time delivery. Quality and delivery are key competitive advantages in this business.

Before using the Strategic Balancing method, the firm had hired a respected globally recognized management consulting firm to analyze the situation. This firm applied standard management techniques. Several factors were used in the analysis, including cost per mile, shareholder value, and the attractiveness of outsourcing versus core competency. The consulting firm recommended closing the trucking operation and hiring an independent provider. The CEO was unsure of the usefulness of the recommendation and subsequently decided to wait on making a change to the trucking function.

Case Study: Quality Foods Incorporated (continued)

Quality Foods then applied the Strategic Balancing method and refined its vision and mission statement. Its new vision became "To remain the country's premier food distributor while initiating growth in the European Union." Its refined mission became "To provide the highest-quality, best-value food distribution in the country."

The trucking analysis was then rerun using the Strategic Balancing method. Since profits are derived from revenues minus costs, to increase profits, Quality Foods needed to raise revenues, decrease costs, or do some combination of the two. Outsourcing the trucking would decrease costs in the near and foreseeable future, resulting from a combination of reduced infrastructure, maintenance, and insurance costs. However, quality of service could not be guaranteed. The company presently enjoyed a premium margin over its competitors because of a reputation for excellence. Excellence in food delivery is closely correlated with on-time delivery, proper temperature, and proper moisture maintenance. The firm's strict adherence to trucking schedules and lease of high-quality, atmosphere-controlled containers (which it calibrated weekly at a tremendous cost) facilitated the reputation. Quality Foods' mission statement did not provide for a low-cost identity; it provided for the best value and highest quality. Therefore, there was no desire to lower margins to gain market share or make other changes that could sacrifice the identity of the firm.

Over the years the trucking function had become part of the company's core competency and was benchmarked successfully against other trucking companies. Trucking was a contributing factor to the reputation for excellence and to Quality Foods' ability to maintain higher margins than its competition.

After applying Strategic Balancing, Quality Foods decided to retain the trucking function. A new campaign was launched to clearly communicate this advantage, both internally and externally. Corporate collateral was changed to reflect this competitive advantage, and the sales force was retrained. These actions increased revenues and, in turn, increased profits and shareholder value.

Furthermore, the operations in the European Union were redesigned to incorporate some of the techniques used in the home-country trucking operation. There were differences in the European Union strategy, however, to account for much higher consumer expectations and the smaller shipments made to local stores. The European Union operations were reevaluated every six months because of their more volatile start-up nature, while the U.S. operations were reevaluated yearly as part of the strategic planning process.

CHAPTER REVIEW

A clear vision (where it is going), mission (what it does), and set of goals that is pervasively communicated within an organization are essential to good decision making and a cohesive management. These statements provide the groundwork for innovation to take root.

Knowing your customer's customer's needs and values and then working backward through your own organizational goals and operations (formally stated as vision, mission, strategies, tactics, and operations) can lead to market-busting insight into the gaps and overlaps of product offerings. With this insight, formal and informal plans and identity statements can be revised while innovative solutions are developed to capitalize on this advanced level of market understanding.

10

PROCESS INNOVATION

"Whether or not the standard of living made possible by mass production and in turn by mass circulation, is supported by and filled with the work of us hucksters, I guess is something that only history can decide."

—***Leo Burnett****, advertising executive, 1891–1971*

This chapter presents the case for process innovation. The relevance and application of process innovation are detailed in several examples, from the classic Henry Ford assembly line to Wal-Mart and McDonald's. There are estimates suggesting that process innovations are responsible for fully one-half of the $2 trillion U.S. economy. Certainly they are worth looking at in detail.

Process innovation is defined as the area of process design that transcends the current state of the art. Henry Ford's work in the automobile industry is the perfect example of process innovation. Henry Ford used process innovation to design and fabricate parts that were interchangeable and easily assembled in a timed fashion on an assembly line. Ford created the assembly line method of fabrication. This method disrupted the established, individually hand-built method of the day. Ford further innovated the assembly line by timing the process so it had a logical flow. Employees performed one operation all day and became experts at their operation. The parts arrived and were assembled to the frame, and the frame grew to become a car—all in one sophisticated, choreographed dance.

Ford innovated vertically in his gigantic factory, the Rouge Plant. This manufacturing plant warrants an entire book alone to describe the intricacies of its functions and the planning that made the plant run efficiently. Amazing processes that developed in the Rouge Plant included harvesting rubber from African trees, converting coal from Pennsylvania to usable fuel, and transporting hundreds of items from around the country and the world all in a seamless integrated process. Ford moved raw materials into one side of his plant, leading them through his various processes, and produced cars and trucks out the other side. Ford even used Midwestern-grown soybeans to make some of the knobs and "plastic" parts inside his trucks and cars.

But let's fast-forward to 1980 and 1990. If someone were to ask what firm has the greatest "process innovation," who would you think of? UPS comes to my mind. UPS is one of the world's largest employers of industrial engineers, all crunching the numbers and making UPS a finely oiled machine. Or perhaps you may be thinking about UPS's competitor, FedEx. Renown for its innovation in IT infrastructures, FedEx is able to track a package in real time. Or maybe you will think of the Human Genome Project—a fascinating process innovation of tying scientists together and linking them to latent supercomputer capacity in the great race to uncover and describe the human genome in detail. Cracking the DNA code was a brilliant process innovation that broke through traditional policies and practices.

The one I am thinking of, however, is such a blockbuster that it literally shaped landscapes. That superiority of market performance is, of course, Wal-Mart. Sam Walton used a combination of modern IT methods, both hardware and software, advanced computer modeling mixed with financial accounting, and market forecasting coupled with a relentless pursuit of supply chain efficiency to produce unmatched performance and margins on every dollar sold. Wal-Mart refined the supply chain to the point where it was taking pennies out of pennies. Most big supply chain efforts and IT efforts, those that succeeded, had only managed to take pennies out of dollars and tens of dollars. Wal-Mart was orders of magnitude better than that. It honed the supply chain to the point where it knew where every box came from, the time the box would arrive at distribution, and the time the box would arrive at a store. Wal-Mart knew how many were on the shelf and how many would be sold that week, that month, and that year. It knew where to stack the product and when to move it. It knew the exact carrying cost, transport cost, and storage costs to corporate and local, it knew when to order more, and it knew when to drop an item. Wal-Mart knew how much to sell it for, the loss leader, the break-even point, and the profit, depending on the product and strategy.

It refined each step individually and then looked at the system overall and refined that. This was very advanced thinking at the time. It is still part of Wal-Mart's core strategy as it sometimes opens over 100 stores a year. Whether you love or hate Wal-Mart, in the area of process innovation, it is king.

My favorite example is not Wal-Mart, however. My favorite example is of Mr. Michael Stadther, a computer scientist turned kids' author. His radical process innovation broke through established industry norms. It broke through societal

CASE STUDY: *A TREASURE'S TROVE*

Stadther's book *A Treasure's Trove* took the single most common, timeless, universal desire of every kid in the industrialized world and made it real. He created a real treasure hunt. Not a kid's story, not an out-of-touch search for the Titanic, but a real, possible, fun mystery that any kid could afford to participate in. The entertainment goes beyond the book; it seeps into their present lives. It is not passive; it is interactive. When he did this he broke through what people thought was possible. He made fantasy real. His idea was to hide real treasure around the country, treasure of significant financial value, and then write a book about it. This is an example of inverse thinking (detailed further in Chapter 14, "Solving the Risk vs. Innovation Dilemma"). He began with the end goal in mind, where his customer (the parent) and his customer's customer (the child) were both satisfied. He may have said to himself, *Here is a real treasure. Now let me write a fun adventure book around finding it. This is what every kid always wanted. Now I'm going to give it to them—not just write about it. And, oh, by the way, it is educational too, so the parents can be involved and support it. I'm not going to make something up. I'm going to do something real.* The difference between real and made up was totally disruptive to the common authoring and publishing mind-set. So contrary, in fact, that he was turned down a dozen times at major publishing houses and ended up self-publishing. Now the work is a huge success, and the book spent several months on the *New York Times* Best Seller list for kids. There are talks of movie deals, sequels, and the like. It is a business success, much to the dismay of those earlier publishers who turned it down.

norms that existed for centuries, and it was also quite novel as a business idea and method. He thought about his customer's customer, and he triumphed.

The self-publishing industry is only a few years old and is rapidly solidifying as a legitimate industry representing a real threat to traditional publishers. The self-publishing industry lacks one key ingredient: oversight. There is no filter, no hurdle, no checks or balances. If you have 500 dollars and a manuscript of any sort, you can make a book, regardless of quality or content. You can make an e-book (an electronic book designed for soft delivery though the Internet) for even less. I envision some day a self-policing independent grading or independent verification and validation (IV&V) mechanism for this industry, helping provide the clout its legitimate authors desperately want. For now, and possibly for the best, the free market provides this natural selection filter. Outlets like Amazon.com list many of these books, and the wise shopper reads the volunteer reviews available. Early adopters, folks who buy on risk with no reviews, are the brave souls that help the rest of us who are more cautious. Bloggers, people who assemble and distribute free information via the Web, are also getting on the book review bandwagon. In this case the process innovations of one industry make the process innovations in another possible. Some are still at the intake stage; others are further along in the combust stage.

Let's pull that thread for a moment. In this current example, we have a cascade of innovations in multiple mediums that make self-publishing work:

1. We have the summary of innovations that make the World Wide Web. Both hardware (switching stations, servers, fiber optics, etc.) and software (browsers, filters, antivirus, e-mail, etc.) combine to make it all work.
2. There are innovations in the personal computer, putting incredible power at an affordable price in almost every home.
3. There are innovations in the business model of Internet hookup, Internet-based money exchange, Web pages, and sales outlets like Amazon.com and eBay, bringing affordable dial-up, digital subscriber line (DSL), and cable access and business capability into almost every home.
4. Then there are the innovations in publishing software that allow a common manuscript written in a common word processing program language to be converted to a read-only, protected format and reduced to a small-footprint file that can be easily delivered via the Web.

All of these work together for a benefit, but they do not necessarily directly enable on-demand publishing, e-book creation, and delivery, nor do they provide an alternative to traditional publishing. They, as a group of technologies, enable all these industry changes. For e-books in particular, they are a critical part of the data delivery and money collection mechanism. But they do not make the book. The real moneymakers for these authors, the factors that are ahead of the services and infrastructure, are the specialized Web sites that deliver the content and collect the money and the specialized software that helps create the content. These are the process innovations in one industry (software) that made the innovations in another (self-publishing) possible. These are the real blockbusters. In this example, a few smart programmers saw the coming tidal wave of authors wanting an outlet beyond traditional publishing, and they responded by writing software specifically for e-book creation. Their task was considered absurd at first. Microsoft Word could easily handle most small manuscripts, and Adobe Publisher could handle the larger ones. These files could then be converted by the industry-standard Adobe Acrobat into Portable Document Format (PDF). However, e-book authors really needed a different product, a product similar to Adobe but with added security, copyright, distribution, and tracking features. These programmers developed comprehensive products to fill that need and have been cleaning up financially ever since. There are about 10 good e-book products on the market now, perhaps 20 total, but the three most popular are the ones that make all the money, on the order of hundreds of thousands of dollars per year. In effect, they enabled an industry by going out in front of the industry.

There are two types of processes: value-added and non-value-added. Most firms have a mix of both, with an unfortunate healthy dose of non-value-added (NVA) activities. Much of my work in the past 20 years has been in the defense industry. Here, in the government- and quasi-government-run oligopolies, NVA lives happily. It is important to note that while the defense industry and government itself have a staggeringly high percentage of NVA compared to other industries, there are some good reasons for that—not the least of which is IV&V. IV&V, mentioned in passing earlier and a form of quality control, checks and balances, requirements management, and contractual fidelity, is a cornerstone of the industry. This industry is spending taxpayer dollars and has a fiduciary, ethical, and moral responsibility to do so with great care and accuracy, which for the most part it does. Those "extra" (in comparison to other industries) checks and balances are warranted, especially on the technical side. Much of what is developed is downright dangerous stuff. It must be carefully engineered so that it does not hurt the operator or innocent civilians nearby. That being said, much of the paperwork, reports, and other items are often redundant and little understood. Some effort to make NVA leaner—to rely on hard-hitting, brief, singular studies and reports for decision making—could help reduce costs over the current multiredundant system.

The point here is not just to improve performance by incremental steps for the sake of additional profits. Many industries compete on slim margins, and a process innovation that yields 3, 5, or 10 percent improvement in a key metric may be the difference between leading the market and bankruptcy.

Think of your own industry and ask yourself where are the NVA steps. Determine if they can be eliminated. Will their elimination give you an incremental advantage or a disruptive advantage? If incremental, is it worth the time and cost? I suggest if you see an incremental advantage to eliminating an NVA, perhaps it is better to keep digging and find the root cause of the activity. Keep asking "why" until you find out how much value the final answer has.

Asking "why" five times will usually get you to the root cause of any issue. Ask "why do we do this process?" Once you have an answer, ask "why" again. (Some consultants call this method "the five whys.") Document each step and then decipher if it is value-added or not. Does it provide direct value to the customer for the money they are spending? If it is a process not linked to the customer, such as an overhead activity, your research should be all the more rigorous. Check to see if the process is still valid. In some cases the answer to "why" may be something like, "Because GAAP—or other government regulation—requires it." When you get to those types of answers, the next question should be this: "Can we refine the process by providing this regulatory element in a more cost-effective way while still meeting the requirement?" Reengineering proponent Michael Hammer taught us "don't automate, obliterate." That mantra of the early 1990s was rooted in the five whys. Look at what you are trying to accomplish and ask, "Does this (thing, process, part, message, outsourced activity) directly contribute to what we are trying to accomplish?" If it does not, ask, "Why not delete it?"

Then get radical and ask, "Can we replace it with a process that leapfrogs us in our industry? Does this process we are trying to eliminate or refine exist because of a repetitive problem that we can solve in a way that provides a solution for sale?"

What about replacing something that is not broken? Suppose there is an established process that works quite well, and suppose your company is also doing well. But suppose also that you are accustomed to innovating throughout the entire life cycle process and see an advantage, however vague, to replacing a tried-and-true, fairly efficient process. To what end? *There are times when innovation itself is the value and a market leader emerges because the buying public was intrigued by the perception and the status quo was not even broken.* Such is the case with the Dunkin' Donuts and Starbucks debit cards. These cards replaced money. Money works pretty well, last I checked. In fact, a consumer needs money to buy the card, and it takes a few minutes to buy or refill your card each time that is needed. How could this possibly be valuable to anyone?

These companies saw the hidden value in shaving seconds off of their customer's morning routine. They saw the convenience of the customer not having to check if he or she has cash on hand. Customers just need their trusty card to have a hassle-free morning. This innovation filled a subtle, hidden need in the customer's life, fixing something that was neither broken nor inefficient, to a great end. In addition, the card acts like a loyalty anchor, bringing the person back in time after time. Furthermore, much like the supermarket loyalty cards, the customer's purchasing habits can be evaluated and other gaps can ultimately be determined. At Starbucks, it has become a sort of identity icon, flashed along with a person's iPod in the checkout lane. The card says, "I am a hip, together person. Are you?" It acts like a catalyst, improving the thrust end of the coffee shop's existing engine.

CHAPTER REVIEW

Think of Wal-Mart, Henry Ford, Michael Stadther, Starbucks, and others. What will it take to be an industry leader with a refined process that will cause demand?

Taking it a step further, a refined process will cause an entire industry to exist where there was none before. This level of industry process leapfrogging is intelligent innovation.

MAKING IT REAL

Some tools to research and apply include the following:

- Process measurement, classic industrial engineering
- Time studies, workload balancing
- Make versus buy/trade studies
- Resource management and balancing
- Material requirements planning (MRP), distribution, and logistics systems improvement
- Whole-life service and support simulation (WLSS) and improvements
- Factory floor automation, modeling, and simulation

In addition, to realize the true benefits of process innovations, be sure to include the following steps:

1. Perform a market survey in your industry.
2. Expand the market survey to include your customer's customer.
3. Perform a gap analysis on the customer's customer, asking if/how their needs are being met by your offerings and by your competitors.
4. Perform a gap analysis on the customer's needs, asking how/if their needs are being met by your offerings and by your competitors' offerings.
5. Combine Steps 3 and 4 to form a simple map of offerings and gaps and needs.
6. Work backward from those needs through your own organization's offerings, looking to fill the gaps and eliminate the overlaps.
7. Leave room for the most seemingly ludicrous suggestions during this process. Perhaps somewhere in the analysis is a profitable, landscape-changing idea.
8. Think of processes that enable filling the gaps. A process innovation can be as powerful or more powerful than an actual product offering, and often they go together.

11

SOUFFLÉS AND COMPUTERS: WHERE PROJECT PLANS MEET INNOVATION

> "It has been my observation that most people get ahead during the time that others waste."
>
> —***Henry Ford***, *founder of Ford Motor Company, 1863–1947*

> "The greater danger for most of us lies not in setting our aim too high and falling short, but in setting our aim too low and achieving our mark."
>
> —***Michelangelo***, *Italian Renaissance artist, 1475–1564*

The tension between planning and execution is a natural, healthy function of most product development activities. Likewise, thorough requirements definition and the detailed project plans that help accomplish those goals often conflict with the need for exponential leaps in design or process required to accomplish broader organizational financial and strategic goals. Mr. Mike Dumas of Newport News Shipbuilding explained an important component of the need for planning and thorough engineering when he said, "Definition (of your product design) by trial and error is expensive and slow." His point is well taken, and yet we know there are times when the unplanned event with no definition (Post-it Notes, microwave ovens, etc.) leads to unplanned breakthroughs. That leads us to dive a little further into management, because without good management, there is no execution of the innovation. It is wasted, or never cultivated at all. In this chapter we discuss an advanced planning technique used successfully to accelerate innovation. It has been used to propel both technology and the people behind it to new levels at a greater speed, reaching market sooner with a better product.

The Strategic Balancing method may be too detailed for some small projects. Or perhaps it is too time-consuming to develop the VMSTO plan for 100 different internal research and development (IR&D) projects. Regardless of the size or complexity, two basic questions must be asked: (1) where is this going? and (2) what do I need to do to get there? Then make a plan. With this in mind, let's look at the next I-Factor:

I-Factor 6: Begin with the end in mind, communicate it to everyone, and set interim milestones that drive you there.

Put the end goal on a scratch pad with a few steps along the way, a date for each step, and any equipment or resources needed for each step. This becomes a plan. A more formal project plan is known as a plan of action and milestones (POAM). POAMs are perhaps best known by their Microsoft Project counterparts. I strongly encourage the use of a Microsoft Project type of program to keep track of, manage, and refine your POAMs.

POAMs rely on milestones and interim milestones. At times, the milestones are fairly evident. At other times, they can be developed only through experience or accident. Milestones can be as simple as the beginning, the middle, and the end. Or they can be as complex as needed to track and propel the project with accountability. POAMs are also used along with other tools such as Program Evaluation and Review Technique (PERT), Graphical Evaluation and Review Technique (GERT), Critical Path Method (CPM), network analysis, and so on. It is not the intent of this book to review and utilize all of them, but instead to show a simple direct method of using a plan to effect a higher innovative yield on time and energy spent.

If a good POAM begins with the end in mind, what is the end? Is the end outcome "innovation"? Is it to have a new bagel shop up and running before the September rush? Is it a technically complex satellite guidance system with performance that exceeds the current state of the art? Is it a corporate merger completed that was previously disallowed by statute? And how do we know when the end is reached? What metrics are involved to get us there?

And what of "innovation?" Do we schedule it as a line item in the POAM, like the famous cartoon of the mad scientists with drawings all over the chalkboard and a little box that says "And a miracle happens here?" In fact, the project plan itself can be an area of innovation. This area, perhaps more than any other, is responsible for huge successes and huge failures of some of the most famous projects in history, including the Hoover Dam and the Empire State Building.

Let's take each element that makes up a project plan separately, first focusing on tracking. Project-tracking milestones are typically linked to basic organizational functions, such as (as absurd as it sounds) the beginning, the middle, and the end. Other typical milestones include, in rough order, the following:

1. Beginning, problem definition
2. Team forming and goal setting
3. Brainstorming session
4. Initial architecture
5. Initial design
6. Final design
7. Final review and integration
8. Completion

Business-oriented milestones, especially those that must be folded into the design or service preparation, can also be included, such as the following:

- Funding allocation
- Market segmentation
- Target pricing
- Focus group results
- Distribution strategy

Propelling milestones (my fancy name for interim milestones) are a little more strategically thought out. They represent key accomplishments that the team must accomplish by a certain date to keep things moving toward the end goal. Interim milestones help drive the group, are good for planning and division of tasks, and so on. A propelling milestone may also be an "interim" milestone, specifically chosen for a schedule-related goal. For example, a propelling milestone like "funding procurement" may be inserted between 2 and 3 in the preceding list, early on in the project. Obtaining the necessary funds normally involves management review of the plan and goals and includes a tacit commitment and understanding by those outside of the project, requirements that will be needed throughout the project. Likewise, it is often useful to insert an artificially created "team review" or "preliminary customer presentation" sometime early on (and recurring) to create a sense of urgency and closure with the team. A good propelling point for a review milestone would be after milestone 4, initial architecture, and again after milestone 5, initial design.

Use a best-case estimate of a potential due date for these types of items and *subtract* a week or two. This puts a sense of urgency on the team, which often leads to creative solutions to difficult problems. This type of schedule pressure can also, if presented correctly, help the team ignore superfluous goals, personal agendas, and areas of nonproductive curiosity. All these distractions are in abundant supply with any process that involves humans.

One of the main findings in the ESRM study suggests that teams under tremendous time pressure actually *outperform* teams that have adequate time in the schedule. The research actually shows that a shortage of schedule, funding, and foreseeable technical solutions was best. I found this phenomenon clearly exemplified in the story of the development of the Ford Taurus, the NASA Pathfinder, the Mazda Miata, and others.

This brings us to the seventh I-Factor:

> **I-Factor 7: An attitude of resourcefulness can overcome aggressive milestones and resource/material shortages.**

Creating an artificial shortage of time and money (if none actually exists) will motivate a team to perform better. Using an interim-propelling milestone, inserted to force some urgency and closure, will create this pressure. It must be chosen carefully to achieve the desired effect without undue stress or absurd closure to incomplete issues. For example, Appendix C shows an actual POAM that was used to bring a satellite navigation system online more than one year earlier than originally planned.

These types of schedule-propelling milestones can also have a serious counter-effect on the psyche of the engineers, designers, and "smart people" doing the work. Many employees by their nature are trained and genetically engineered to thoroughly complete their work. They do not want to make "guesses" and "projections"; they want to make final decisions based on complete research, analysis, and experimental results. This desire results in a natural and healthy-if-managed tension between the higher-risk exponential leaps into the "educated guess" camp and the "detailed decision with proof" camp. Simple physics state that the latter camp takes longer to get to the end than the first. So, in a schedule-constrained project, the manager often has to coax the "best-possible current decision" out of the team and keep them going forward, especially in the early conceptual phases. Some consultants call this the 80 percent solution. Of course, there is risk here: Starting off in the wrong direction early on can lead to devastating failure. Some project managers will choose not to do this. Some project managers will choose to use parallel paths and let the team explore two or more early concepts under equal pressure. Some project managers will split the team and let competition win. Finally, some project managers will put "primary" resources on the leading idea and let the trailing idea be worked by the backup team, and so on.

Develop interim milestones that suit long-term and short-term project goals. Ultimately, the practitioner will be able to develop targeted, strategic (innovative) interim milestones. A POAM can be as detailed or as high level as desired. In large, complex projects, POAMs may be three or four levels deep. Each level contains the information necessary for a particular group to enact its operational responsibilities.

Behind every POAM (or the much less formal napkin plan or business plan) is a "build strategy," whether known or unknown. A build strategy is the order of events that the project manager, the program manager, the inventor, the innovator, or even the customer uses to accomplish the task. That order of events can be purposed, thought out, and strategically planned for efficiency of time, physical resources, and knowledge resources, or it can be unplanned.

While the build strategy is seldom discussed and even less understood, it is critical to success. Let's start with a simple example. Then we will use a more complex example to show where the innovative aspect influences the outcome.

A simple example of build strategy is found in the making of a soufflé. A soufflé depends on the precise whipping of the eggs and the assembly of all ingredients in an exact order to become a good product. Any deviation in the order, timing, or consistency of raw ingredients or vigor of the whipping will result in disaster. A cook orders all the ingredients first, works through the recipe, and assembles the masterpiece. With the possible exception of the first person to invent a soufflé, there is little innovation here, although there is a very precise and critical build strategy.

Now think of something more complex: a laptop computer. Several components, such as the screen, keyboard, case, power supply, and battery, are made by different companies in different parts of the world. The laptop is assembled in Taiwan and shipped to the United States in a timed, orchestrated symphony of coordination. The build strategy for a laptop resembles the build strategy for the soufflé. The keyboard must arrive first in order to be attached to the case. Then the screen is snapped in and connected to the power supply, which is connected to the battery, and so on. All of these events are scheduled by an automated pull-driven supply chain management system, and any supplier hiccup results in stopping the line and a drastic increase in holding costs, endangerment of customer relations, and so forth.

Again, we can all picture this system and understand its relative complexity and importance. But what of innovation? Well, there was probably some innovation required by the industrial engineers to balance the line. Perhaps the power supply and battery suppliers were across the globe. Power supplies were made in China, and the new lithium ion batteries were made in Texas. The case was stamped in China of titanium made in Russia on equipment made in Sweden. The schedules would not coordinate at first, and the engineers had to find unique ways to stock the earlier-arriving parts while accelerating shipment of the later-arriving parts. The businesspeople had to find new ways to construct the contracts so that there were incentives (and penalties) for on-time delivery, while the carrying costs of the earlier suppliers were absorbed in the overall cost structure. These process and scheduling and business innovations are a key part of the overall success of the project, and in fact, the laptop could not have come to market at all (for the correct price–quality point to match its performance and target market) if not for these unsung, behind-the-scenes innovations. But ultimately it is the build strategy that is key here; it is the sheet music to the symphony of arrivals and assembly.

The next level of sophistication beyond POAMs and milestones is optimization. Several software programs and numerical methods techniques allow the program manager to simulate results of a given path. The simulation can focus on process or POAM design or go further and look at optimizing and validating the plan. On a very complex system that relies on hundreds or thousands of systems and subsystems, such as the development of a satellite or ship, this next optimization step using software is essential. Some numerical methods include color Petri nets, system dynamics modeling, enterprise resource planning, network analysis, multivariable testing (MVT), and so on.

DISCUSSION QUESTIONS

1. Can you think issues similar to those in this chapter in your own firm?
2. Are there areas where your processes are stuck or limiting?
3. Are there ways your build strategy can be improved?
4. Are there opportunities for parallel processing or the use of interim (propelling) milestones?

Ultimately we are talking about constraint management: the ability to balance and even take advantage of constraints and to organize them in such a way that they become assets, not difficulties. This is another area where your mission and vision statements come into play. Without a good vision and mission, it is difficult to craft a good POAM and nearly impossible to justify your choices as well. However, a good POAM is much more than a plan to get you to the end, whatever that may be. A good POAM also takes into account constraints, capacities, and work content. A really good, innovative POAM takes this process one step further and transforms constraints into opportunities. This ability is rare, and it is the Holy Grail of constraint management—even of management itself.

CHAPTER REVIEW

Plans that begin with the end in mind are an important contributor to the overall success of projects. Project management that includes interim propelling milestones and celebrates a slight shortage of time, money, or technical maturity often succeeds well beyond its original goals and beyond related projects that were managed without limitations or stretch goals.

MAKING IT REAL

Use the following steps to help your company realize optimal project management:

1. List the project's goals in terms of financial, schedule, and technical performance. List any dependencies or linkages on these goals.
2. Prepare a rough plan of actions and milestones—a POAM. If a milestone is related to a particular goal, put that in the notes for the team to see.
3. Add elements of the project goals into the POAM, providing early and periodic peer reviews of the financial, schedule, and technical performance. Flesh out as much detail of the individual steps behind each milestone as possible.
4. Add interim—stretch or propelling—milestones into the POAM to create decision points that pull rather than push the project through to conclusion.
5. Save the baseline and don't be afraid to revisit, revise, improve, and status-check often.
6. Keep metrics and use them as a management tool; however, remember that there is no metric for "innovation" and you cannot schedule a miracle to occur on "line 37 of the POAM." Although a comparison to the original baseline metrics is useful, it is normal for plans to evolve as more information becomes available. A more useful comparison is made against the actual goals of the project and how they progressed over time.

12

TURNING LIMITATIONS INTO OPPORTUNITY

"Typical. Just when you're getting ahead, someone changes the odds."
—***MacGyver***, *"Pegasus" episode, 1985, American Broadcasting Corporation*

Turning limitations into opportunities is the single greatest factor that the 1,000 successful projects in the ESRM Survey had in common, and it is the subject of this chapter. Data from the survey showed that projects having limited or severely limited resources generally met with success. Limitations can come in many forms, both internally and externally. Internal examples include available employee labor, finances, machine time, and the like. External limitations can be more frustrating and are often more uncontrollable. These include regulatory roadblocks, working capital interest rates, shipping costs, technology swings, and so on. They can make or break entire industries. Understanding capacity and resource management is an important aspect of doing well in this continuous, life cycle process called business.

SURVIVING VS. THRIVING: INNOVATION IN THE MIDST OF LIMITATIONS

Finding innovative ways to work around limitations can mean survival, where the thrust of the engine is just enough to keep moving forward and where the fuel intake is just slightly less than the thrust output. Some situations are so dire or so complex that managing this level of thrust is quite an accomplishment and should be considered a huge win, even though the organization is barely surviving. That was the case with some real-estate developers in the mid-1990s in the eastern United States. Some multimillion-dollar developers were reduced to bankruptcy in a matter of months, as dot-com and biomedical organizations failed or downsized in droves, leaving a glut of office space. Conditions in this tough market were exacerbated by the banking community, whose loans for both working capital and mortgages became increasingly difficult to acquire. Some real-estate developers turned to innovative solutions. They started finding partners, teaming up with other developers on deals and providing full-service brokerage and build-out project management. A few real-estate developers found niche survival by catering to the downsizing firms in popular accessible suburban locations in Massachusetts and the technology corridor in the District of Columbia.

A few firms, however, went much further. Some real-estate firms looked at the market downturn as an opportunity. Making money during a bear cycle is a little trickier than during the bull cycles. Finding innovative ways to capitalize on limitations, to harness them instead of fighting them, and to use them to your advantage is Intelligent Innovation. These firms made millions by buying out distressed properties, often leasing them back to the original owner in the same day. This was a simple exchange-of-debt service and ownership similar to the Rockefeller model some 80 years earlier. Rockefeller was a counterintuitive buyer. He bought when everyone else was selling, and he sold when everyone else was buying—that is, buy low, sell high.

I often think of my grandfather when I hear examples like this. He was in the first group, the survivors. Before the Great Depression, he owned several apartment buildings and houses and three small, family businesses. Not bad for a guy who barely spoke English, could not read or write, and nearly starved to death on the streets of New York the first two years after landing here from Europe. He was shrewd, smart, hardworking, and opportunistic. But he was overleveraged. When the Depression came, he lost his buildings one by one. They went back to the bank, which had no interest in them and often sold them for pennies on the dollar. By the end of the Depression, he had two small stores and his own house left. His family was still better off than most, and my grandmother spent many days per week making soup for the poor in her neighborhood from whatever food was about to go bad in their grocery store. My grandfather survived but never really thrived after that disruption. Some other entrepreneur owned his buildings and houses.

Was my grandfather particularly innovative, either in his buildup or in his survival? I don't know. I tend to think his methods were more commonsense and resourceful than new, novel, or innovative. Indeed, by the end, he was taking cars,

CASE STUDY: ENCORE ACQUISITION

The company Encore Acquisition, run by Jon and Jonny Brumley, a father–son outfit, turned a seemingly problematic situation into an opportunity. The basic premise of Encore is that it bids on and continues to buy aging oil wells in Texas. Texas has long been overshadowed by the supply of oil that comes from South America, Africa, Europe, and the Middle East, but the old wells in Texas and elsewhere often still have up to 70 percent of their energy remaining. The oil is just harder to get out, because the "sweet oil," as it's called, has already been pumped off the top. An oversimplified explanation is that the oil remaining is too thick to easily pump out with a traditional process. It is down there, but it is "stuck." It requires an investment of technology, energy, and time that the big oil companies—usually the former owners of these wells—are no longer interested in. The term "energy" in the last sentence refers to organizational resources, people, and time and effort to figure out the unique technological solution each well requires, as well as, more literally, the electric and hydrocarbon energy required to get this heavy oil up and out.

Many of the methods to remove this oil require energy to change its viscosity to the point where it will pump. This creates an energy-return equation that borders on unfavorable. So an energy return on investment—known as an EROI—of 2:1 or 3:1 may be undesirable for the big-name oil companies but workable for a smaller, less burdened company like Encore. In fact, to an outsider, it looks like a symbiotic relationship. The large oil companies need a much higher EROI in order to pay for the expensive original exploration and big deals that they make all around the globe and still meet market expectations for profitability. When a well goes dry and no longer produces at the minimum EROI rate, such companies can sell it to a much smaller, more efficient (for this purpose) organization and still realize a final return from the well. Meanwhile, Encore and others like it are specifically set up to capitalize on these wells and make a profit doing so.

horses, and jewelry in trade for rent and quickly turning it around for whatever cash he could get to pay the mortgages and electric bills.

On the other hand, some companies and people thrive in these conflicted, depressed environments. *They see tiny (or large) gaps in the business model of the day and walk through them, meeting a huge need on the other side of the gap and reaping the benefits.* These methods can come in the form of a monumental, competition-crushing strategy that changes the fate of organizations, providing thrust well beyond the input. These opportunities are few and far between, but those who recognize them for what they are and forge ahead while others cower are the people and companies we read about in *Forbes* and the *Wall Street Journal.*

There is another example that is more familiar to most and shows that innovation in process can even overcome huge barriers to entry. In 1971 a few people got together with the audacious idea that they could start an overnight shipping company. In a world already filled with established players like the U.S. Post Office, UPS, and to a lesser extent DHL Express, it was an uphill battle at the least. This team, lead by Frederick W. Smith, saw things differently. Yes, there were solid competitors who were efficient at their game, but there was still opportunity. There were holes in the way these companies did business, particularly in the delivery of urgent packages by air. Each hole was an opportunity to do things differently and more efficiently and beat them at their own game. The people at Federal Express (now officially just FedEx) saw a gap to be exploited with a huge market on the other side, to the tune of $31 billion, according to the FedEx Web site.

The limitations were tremendous. There were natural barriers to entry, such as name recognition and consumer outlets; at the time, everyone knew UPS and where to find a UPS counter. There were regulatory barriers to entry, with laws and approvals required, and there were capacity constraints to be overcome, with airline and trucking hubs to be leased or built or approved. Anyone without the vision would think this undertaking was audacious, even ridiculous. But in this case, the core team of FedEx had a vision and mission based on a solid business plan, along with the innovative tactics and operations to back it up.

By starting its logistics design with a clean slate and modern-day information technology, the management team at FedEx developed a new business model and was so successful that its name became a verb meaning to ship a package overnight, regardless of the actual carrier. Its Intelligent Innovations led to a multimillion-dollar return on investment.

When we are discussing limitations and constraints, it is helpful to review a couple of basic work measurement terms: work content and capacity. *Work content* is composed of two components: hours and tasks (or skill). Work content is the amount of effort (typically measured in worker-hours) required to accomplish a task. Knowing the work content required for each task or each portion of a project or program is a fundamental key to performing good, basic planning. It also impacts advanced parallel processing and other innovative management techniques. This relationship is similar to the engine control unit detecting the fuel flow into and out of the injectors. The information is compared with the required fuel flow for a certain speed setting. The ECU balances supply and demand.

Unfortunately, knowing the work content for each task, big or small, can be a very tall order. It is a bit like trying to find a word in a dictionary when you don't know how to spell it. The product development process, especially the early phase, has no known or set times for each activity. Most business activities are somewhat variable in their length of time to complete; some are even random. When there is no historical data to project, you will have to become comfortable with guessing the activity and its time requirement. POAMs are very iterative and must be updated each time more current information or better guesses are available. Again, that is why software is handy.

Capacity, like the work it accomplishes, has two elements: time and skill. A mismatch between work content and capacity is often at the root of most complaints from employees, managers, and customers. In Massachusetts, there is an old saying that refers to the tangled spider web of cow paths now called roads: "You can't get there from here." Well, that's how most people feel about the overwhelming amount of work they have each day. They can't get to there, to the finish line, from where they are currently. That's how most project managers feel about the daunting project plan ahead of them when viewing the four tired engineers they have at their disposal, when the project plan says they need 40. Obviously, capacity planning is tightly linked to POAMs: Any opportunity requires action to bring it to a logical conclusion and action requires capacity. Capacity includes people, money, energy, materials, and even governmental permissions and partnerships.

I would not have included these several chapters on capacity planning if not for the glaring historical content linking it to successful business equations. Capacity planning is linked to innovative methods and approaches to circumvent capacity-related shortages or blockages. In example after example, data point after data point, product development successes have one, consistent, clear, industry-independent, economy-independent conclusion:

> *Limitations are not always a curse. Constraints of people, process, time, and money can be a good thing, counter to all logic, in some situations when solid and creative planning methods are applied.*

The next I-Factor is as follows:

> **I-Factor 8: Turn limitations into opportunities and understand your organization's capacity so that an innovative approach to the opportunity can be capitalized on.**

This I-Factor could easily have some corollaries, such as the following:

- Outflank your competition, who is experiencing the same limitation.
- Eliminate the constraint by redesign of the product or system.
- Make use of the government regulation in your business model.
- Add an optional service offering (that uses the same infrastructure).
- Leap ahead of the market and create demand for something only you have.
- Enter low-margin businesses where others fear to walk.

I-Factor 8 occurs at every step along the turbine engine process. At the intake, it appears in the early funding of R&D. In the compressor, it correlates to the allocation of people and test equipment (or partnerships), who take those early concepts and flesh them out. In the engine, it correlates to the metering of the fuel to provide the exact level of combustion (output, quality, packaging, warranty) that the market demands at that price point. In the turbine stage, it correlates to the delivery strategy, distribution partnerships, and margin management that either win the day or fail all previous efforts.

Corporate goals and cultural values heavily influence the capacity planning strategy. Consider priorities and balance. One successful firm dictated that all employees get 15 percent of their weekly time as "play time"—time to innovate, think, or experiment on any nondirect activity that may someday benefit the company. Today it is a very successful company with several lucrative spin-offs. Typically, capacity is calculated as eights hours per day per person, minus lunch and two 15-minute breaks, morning and afternoon. This number is then multiplied by a fudge factor, unique to each industry, to further reduce it. Machine capacity follows a similar calculation, with downtime for maintenance and repair:

Capacity = Work that can be accomplished in units of hours

If capacity is greater than or less than the work planned, there are only three choices:

1. Move work planned as needed by calendar, balancing the equation on a weekly or monthly basis. For a short-term project, you may want to do this weekly. Each group member should practice doing this.
2. Add or subtract capacity.
3. Add or subtract the work in the queue (this may mean adjusting goals or expectations).

Finally, as you may have guessed, a build strategy for capacity balancing develops. Since certain skills may be in short capacity during, say, week 3, the group may try to move them out to week 7, where they are in low demand. This is an appropriate solution as long as tasks in weeks 4, 5, and 6 are not dependent (serial) on the accomplishment of the task in week 3. Parallel accomplishment is key to success in a tight schedule. The group must decide what can be done in parallel and what must be done in serial. That leads us back to milestones.

CHAPTER REVIEW

Turning limitations, no matter how great or how small, into opportunities is a basic tenet of Intelligent Innovation. Often, great return is hidden behind a limitation hindering an entire industry. Only the brave, resourceful, and opportunistic people will go through or around the limitation to claim the prize. Work content and capacity planning are intricately tied to an organization's ability to capitalize on innovations of any sort and must be adequately understood for success.

Appendix D contains more information on work content and capacity for those who desire to understand these areas more fully.

13

PARALLEL PROCESSING: A STRATEGY FOR SUCCESS

> "And while the law of competition may be sometimes hard for the individual, it is best for the race, because it ensures the survival of the fittest in every department."
>
> —***Andrew Carnegie***, *industrialist and philanthropist, 1835–1919*

Constraints and limitations are opportunities for process innovation. Developing methods to capture and exploit those opportunities is the subject of this chapter. To capitalize on the total formula for innovation, we will end the four-chapter focus on management and planning with a focus on the advanced topic of parallel processing and strategic planning.

Before we discuss parallel processing, it may be helpful to review some elements that are critical to Intelligent Innovation so far:

- Innovation needs to be cultivated, encouraged to take root, nurtured and protected, acted upon, integrated, and capitalized on in every phase of your business.
- Good decisions are the foundation for enacting innovation. They are based on solid vision and mission statements, which provide the basis for relevancy.
- These vision and mission statements determine priorities, strategy, and timing.
- Priorities and strategy are reflected in the overall plan and in the interim milestones.

- Interim milestones can be used to accelerate the innovative process.
- The strategic plan and interim milestones reflect what is worthy in the form of resource allocation.
- Innovative solutions help overcome the competition and constraints via implementation of the plan.
- Understanding your customer's customer is a powerful concept.
- Gaps or limitations in the business model of the day may be hidden opportunities.

To capitalize quickly on these factors, a radical approach to scheduling and allocation is required. Parallel processing is a robust method for overcoming many schedule and capacity constraints so that new products are brought to market ahead of the competition. Parallel processing is a known, repeatable process. Its outcome can be hard products, service offerings, business methods, or a combination of all three. Remember that most of your competitors probably have similar problems. The person who figures out how to capitalize on them—not just survive—is the one who wins.

Dell computer is a good example of how parallel processing can provide an innovative edge. For the past decade, the entire personal computer industry has been faced with declining margins, an increasing need to speed products to market, rapidly changing technology, and intensely competitive market share issues. Dell used a proactive approach to prepare for and embrace an industry with changing demographics. Dell developed processes, partnerships, and distribution functions that help it thrive in a very tough industry. This occurred in part because management asked "how can I work this XYZ situation?" rather than "how can I avoid it?" This approach helped Dell develop a dual parallel distribution strategy, using both direct Internet sales and point-of-purchase kiosk sales to increase overall volume. It is a question of attitude and approach.

Many of the world's most successful firms got their start because of a process innovation, not a product innovation. As we discussed earlier, for Ford it wasn't the car; it was the processes to make the car (and resulting market stratification, price point, etc.). The same holds true for Dell computer, Wal-Mart, and others. These firms relied on multiple process innovations for their success. However, another level of process innovation combines some of the theory on constraint management and POAMs discussed earlier: the parallel process.

In the case study, two methods were used to shorten the plan: constraint-based optimization and the parallel processing method. In the first method, POAM innovation can be obtained by systematically revising the plan, holding everything constant, and then eliminating cost, schedule, technical, or resource constraints, one during each iteration. The POAM is then iterated and summed up, and the

CASE STUDY: SATELLITE LAUNCHES AND THE INNOVATION OF PARALLEL PROCESSING

The following is taken from a real case study, with the industry and name of the company changed to protect confidential information. Satellite Launches Inc. manufactures a key component of the satellite system used for televising major sporting events. This system was under development by the firm, a small start-up in California, for several years. The firm was going bankrupt until a venture capital firm injected $20 million of much-needed capital. The venture capital firm, as is customary, also initiated several aggressive management and financial targets for the firm and its employees.

The firm's primary product was scheduled to reach market in four years. Its first operational tests had been planned two years earlier. In January the customer called and said it needed to drastically accelerate development of the project. Operational prototypes were needed for testing within six months, and full production was to start four months after that. The program manager in charge called her team together and asked if this acceleration was possible.

Most of the team, which included seasoned, talented engineers, physicists, and programmers, concluded that there was no way they could work any faster or harder. They also concluded that the system still relied on several cutting-edge components from the optics, electronics, and material sciences industries that had not even been fully developed yet. Some were even still in the theoretical stage!

Indeed, the system had been conceptually designed with components that were years away from fielding. They were still ideas in scientists' heads, and yet the customer wanted one now.

The program manager realized a simple truth could cut half the time from the schedule: Work in parallel, not serially—the most basic process innovation. She also realized she would have to "manage in parallel" to support it. Most of the people working on the project were scientists and engineers, all trained to work in serial. Item A gets completed prior to item B, which fits into item C. Instead of that linear approach, she reworked the POAM using the parallel development approach and got a 100 percent improvement. Unfortunately, this still did not meet customer's challenge, and more reduction was required. The program manager now knew that she could get further gains by eliminating some high-risk items. She had been using an advanced form of risk management known as three-dimensional risk management, which highlighted risk items by phase. She could accelerate the project by looking at the early-phase risk items and replacing them with items that had similar performance but were in an area of more mature development. The POAM was completely rewritten again. Any task not directly dependent on another was brought forward for a parallel start. It was September, and product needed to ship by March. Resource constraints were ignored for now, and only physical constraints were allowed to keep an item in serial on the POAM.

She showed the new plan to the team, and after the initial "are you crazy?" comments, the team settled down to look at the POAM in earnest. She had done well and only a few changes were made—items that were truly required to be in serial.

The ultimate schedule was shortened by 17 months, and the project was delivered in May, to everyone's amazement. In 2004 the project was a common sight on the evening news, and only a few people in the world really know who put it all together and how they did it way ahead of schedule.

critical constraint is located. Once located, this constraint is attacked with more resources, more money, or more time, or, optimally, it is eliminated entirely.

In the second method, the program manager lines up all of the operations in serial. Next, the entire team lists dependencies below each operation. The program manager then challenges each dependency. The team will find that many if not most dependencies exist "because we always did it that way" or "because we should do that first" but otherwise have little real justification. Dependencies are slashed (with a special eye toward physical and mission safety, which should be given special treatment), and a new POAM is developed where many processes occur in parallel. If two or more potentially parallel processes are owned by the same resource, a decision can be made to either increase the capacity (second shift, outsource, etc.) or put the processes in serial.

Depending on the needs of the organization, the POAM that is most appropriate is chosen. This assumes that the risks in the area that is held stationary and the risks in the area that is allowed to move are considered and mitigated as needed.

Additionally, the softer side of motivation should not be ignored. Charlie Schwab, Andrew Carnegie's top steel mill manager, got an 83 percent improvement simply by motivating his people with a piece of chalk. Schwab walked into a particularly poor-performing steel plant at the end of the shift and asked how many heats were made. The answer was six, the maximum they always made on each shift—no more, no less. Mr. Schwab was a student of people and motivation. He took out a big piece of white chalk from his pocket and wrote the number 6 on the ground where the workers entered the building. He said nothing.

The workers for the second shift came in, saw the 6, and said, "What's that?" Word got around that it was the number of heats made by the day shift. The rest of the process was human nature: competition. The night shift made seven heats, more than they had ever made. The next morning on their way out, they met Mr. Schwab in the entryway. They were all very proud of themselves for making a new record. Mr. Schwab just smiled, crossed out the big 6, and wrote a 7.

The competition went on and on, ultimately stabilizing at around 11 heats per day. With a piece of chalk Charles Schwab had increased output over 80 percent. Behind the scenes of Carnegie's steel mill were many complementary improvements, such as process innovations, product changes, more focused workers with a common goal, better teamwork, and improved communications between managers and workers. It is this overall combination that defines Intelligent Innovation.

A more subtle force is also at play here as well: workplace culture. Fostering a culture of innovation and opportunistic management is often at the root of great successes in all types of organizations. "One of the easiest ways to get things done," Schwab explained to the mill manager, "is to stimulate friendly rivalry. I don't mean in a money-getting way but in the desire to excel." More on this topic of culture, motivation, and attitude is found in the final chapter.

CHAPTER REVIEW

Parallel processing is an Intelligent Innovation concept. It can help an organization drastically accelerate certain development processes, shortening the time for ideas to go through the intake, compress, and combust phases. Additionally, sometimes good old internal competition and goal setting can improve performance and thrust.

14

SOLVING THE RISK VS. INNOVATION DILEMMA

"I like people who do something, not the safe man who stays home."
—***President Theodore Roosevelt****, February 12, 1908, in Times Square at the start of the New York to Paris automobile race*

"The failure is the man who stays down when he falls."
—***David Dunbar Buick***

Risk and innovation go hand in hand. Solving the risk versus innovation dilemma is the key to supercharging innovative ideas and performance. The 1908 New York to Paris auto race was about both innovation, determination and risk—financial, technical, schedule, temporal, and personal. Those brave men and women who trekked over 21,000 miles across three continents and six countries, crossing mountains and deserts and enduring every environmental extreme in immature unproven technology, were far from home indeed. This chapter explores a method and thought paradigm that can be used in almost any situation to quickly weigh risks and returns and make appropriate goal-based decisions, even when considering immature technologies or ideas early on in the intake phase. This brings us to our ninth I-Factor:

I-Factor 9: Understanding risk organizationally and strategically, by phase, can promote ingenuity and performance.

Corollary to I-Factor 9: The risk of not taking risks often outweighs the perceived cost of a potential failure, especially in today's fast-paced global economy.

Consider the following from Robert Ludlum's novel *The Parsifal Mosaic* (Random House, 1982):

> *Undersecretary of State Arthur Pierce . . . breathed deeply, steadily imposing a calm over himself as he had learned to do . . . Whenever a crisis called for swift dangerous decisions; he knew full well the consequences of failure. That, of course, was the strength of men like him; they were not afraid to fail. They understood that the great accomplishments in history demanded the greatest risks; that indeed, history itself was shaped by the boldness not only of collective action but of individual initiative.*
>
> *Those who panicked at the thought of failure, who did not act with clarity and determination when the moments of crisis were upon them, deserved the limitations to which their fears committed them.*

Robert Ludlum's words are powerful, and one could almost teach an entire course on new product development from that passage alone. It helps explain individual efforts in the context of team efforts; it helps explain the necessity of risk in relationship to reward and the concept of opportunity costs in relation to failure costs. In this chapter we touch on all of these concepts, with a primary emphasis on understanding the risk-versus-innovation-return dilemma.

Initially, innovative ideas or approaches are equated to risks. This is a gross oversimplification used only for understanding the initial concept. More complex and more powerful concepts such as inverse risk and three-dimensional thought are introduced later in the chapter. The three-dimensional risk method correlates to all four stages of the engine, mapping out the appropriate risk tolerance for each stage for the particular goals of the organization. The three-dimensional risk method ensures that the winning projects are identified and not eliminated from critical funding and resources too early, while it also ensures that strict schedule or cost goals of mature programs are followed. It ultimately provides thrust to the organization as a whole as the projects mature. The three-dimensional risk method further suggests that those projects that fail or are taken off line anywhere in the process provide value in and of their failure.

Traditional risk management and other program control measures such as earned value management (EVM) are inappropriate for many programs—particularly small projects, technology development and insertion programs, and R&D projects with high yield or fast return goals. A lack of *risk management, or a lack of understanding risk organizationally and managerially, can also hinder ingenuity and promote mediocrity in larger programs. This is especially true during the critical concept exploration and preliminary design phases.*

Intelligent Innovation requires both an understanding of the nature of risk in product development and application of a modified form of risk management. This modification allows innovation to take root when and where it is appropriate. To make this modification, management must link the perceived risk of a project with the strategic goals of the organization and the metrics used to accomplish those goals at a point in time. A new type of risk management evolves from this connection. This process of thinking fosters good program management and innovation without sacrificing robust design.

This new form of risk management, called three-dimensional risk management (3DRM), links each point on a likelihood/consequence curve to the organization's goals, tactics, and strategies. A program phase adds a third dimension to the traditional dimensions of consequence and likelihood. In addition, a powerful graphical summary view is presented that highlights the difference between the risk level and expected return of each item. Taken together, they help foster "intelligent" risk, where the right amount of risk is accepted for each stage of maturity. Having a measured risk raises the overall quantity and relevance of innovation in any project.

Traditional risk management has added value in many programs, providing early warning and correction to potentially catastrophic situations. Rockwell Collins and other major aerospace and defense firms have documented cases reflecting significant schedule and budget savings attributable to their risk management programs. In one case, the company attributed a 17 percent savings to its use of an established risk management program. However, several respondents in the SRM study wrote in notes indicating that traditional risk management had done untold harm to their companies and programs. These programs could have achieved greater goals with less constrained innovative thinking and approaches. Instead, the programs were stunted by the risk management measurement process and grading techniques. Likewise, Charles Holland and David Cochran's book *Breakthrough Business Results with MVT* (Wiley, 2005; MVT stands for multivariable testing) suggests that most business ideas do not work well. Their research indicates that 75 percent of business improvement ideas and methods do not improve results. Some 53 percent make little or no difference and waste resources, and 22 percent actually have a negative effect on results. Innovation that is balanced with the high opportunity cost of managed processes and managed risk is greatly needed. Managers must have a repeatable and reportable process that does not cause an organization to sacrifice opportunities because of an ill-applied risk metric or, more likely, an ill-applied fundamental understanding of risk. *3DRM provides an equation that satisfies both the need to innovate without constraint and the need to march to logical management conditions and methods.*

The following paradox of modern management explains the inherent conflict within risk-managed decisions:

According to research conducted by Greg Stevens and James Burley that appeared in *Research and Technology Management* 40(3), May–June 1997, pp. 16–27, during the mid-1980s through the 1990s, information infrastructures and program management tools and methods were supposedly improving, while the performance of some programs was decreasing. Many programs in both the Department of Defense (DoD) and the commercial sector required constant budget and schedule extensions, while technical performance requirements were repeatedly relaxed. This sharply contrasts with the program development efforts of the mid-1950s through the late 1970s, where "old-fashioned" program management and product design tools (and primitive computers) were used with excellent results. Since the complexity of past programs relative to current programs (given the resources available at the time) can be considered equal, a more in-depth account of the apparent failure of modern management was necessary.

The SRM study revealed that many successful new product development programs showed a high correlation among some management elements, while many unsuccessful or marginally successful programs showed a high correlation among other management elements. As mentioned, the SRM Survey ranked hundreds of programs for criteria such as innovativeness, market and technical success, management drivers, management structure, technical difficulty, organizational culture, schedule and budget adequacy, and ideological urgency.

The SRM Survey data produce six clear but somewhat counterintuitive conclusions. I reiterate them here for clarity and to propose the basis for the somewhat revolutionary and counter-intuitive and countercultural aspects of 3DRM.

1. Many successful programs had inadequate schedule and budget and almost always began with technical requirements beyond the current state of the art.
2. Many unsuccessful and marginally successful programs had adequate budget and schedule or were managed using accepted forms of program management including EVM and risk management (RM).
3. Organizational culture and urgency almost always matter; however, organizational structure (flat or hierarchical) and specific management tools used almost never matter when rating success.
4. Successful programs had a clear ideological mission and usually an equally strong vision or were a known contributing aspect to a clear vision.
5. Successful programs had visionary leadership.

6. Schedule, budget, and technical constraints coupled with (or countered by) an innovative organizational culture and a sense of ideological urgency among the design team have a high correlation with success.

These results are startling. As mentioned in conclusion 6, schedule, budget, and technical constraints, along with an organizational culture that is innovative and has ideological urgency, are often successful. Conversely, adequate budget and schedule coupled with a risk-averse organizational culture and/or a lack of urgency have a high correlation with mediocre performance (both financial and technical).

Oddly, the use of "good" program management including EVM and RM showed little correlation with either success or failure and possibly more correlation with failure. Projects under financial and technical pressure often succeeded despite common understanding that they would fail. These conclusions required further investigation into several accepted aspects of program management, with special focus on the early program phase where, according to Ronald N. Kostoff, "larger steps into the unknown are needed to attain the... objectives" ("Identifying Research Program Technical Risk," *Research Technology Management* 40, May–June 1997).

The SRM written comments suggested that during early phases of a program, the funnel and the compressor, RM professionals had a tendency to overassess the risk associated with nearly every aspect of the program. I have personally observed this over the years in all types of programs, companies, and industries. People are scared by the high risk of the early phases. These poor ratings contribute to early termination of some high-potential programs, marginalization of strategic intent in others, and micromanagement in still others. Ultimately, incorrect risk ratings actually yield lower cost and schedule performance part of the time and a mismatch between program milestones and organizational strategic needs most of the time. *Many risk programs run counter to the very goals they are trying to improve. They can increase total organizational risk and decrease innovation and return by removing necessary risk from a firm's total product portfolio.* These firms wake up one day from their riskless slumber, and their more innovative competitors are filling the shelves with the next round of products to which they have no response. The incorrect risk assessments and downward pressure on results are a direct result of the misunderstanding of the mathematical and cultural background of risk management.

The goal is to improve risk management methods and return on innovation investment (R&D, IR&D, science and technology, or S&T) while preserving the good and necessary early warning provided by risk management and other program management tools. Some of the downward pressure on performance can be reversed by simply educating management professionals about concepts such as "risk-averse cultures," opportunity costs of not taking risks, and the need for innovation at all phases.

One easy step is to further educate project and program managers and foster a constant tension between the structured and unstructured worlds. The structured processes give an increase in efficiency, quality, and thoroughness (EVM, RM, configuration management—or CM, etc.). The unstructured processes give us undefined, exponential, accidentally creative breakthroughs beyond what was planned.

The key that exists in between these extremes is an organization's ability to follow through, to leverage the correct breakthrough ideas (using defined, efficient standard processes) and yield returns in profit (or performance). This ability is predicated on identifying breakthrough ideas correctly, assessing which are supported by (or support) the organization's strategic intent, and executing them while remaining inside the organization's risk tolerance.

Most managers understand this concept but are unable to come to grips with actually managing it day to day. This inability to balance risk and return and to understand the varying success metrics by phase is the crux of the problem. Many firms and even projects within firms are unable to provide a return on investment, manage risk, and harvest the social (public relations), marketing, and technical benefits of breakthrough innovation simultaneously. Usually at least one of the three is sacrificed.

Most organizations do not understand their "risk tolerance" and often apply one tolerance boundary to all programs within the organization, regardless of the specific nature of that particular effort. The real goal is for all efforts (from mainline production to R&D to primary research) to provide returns within an aggregate total acceptable level of financial, schedule, and technical risk appropriate for that phase. Applying a single tacit risk tolerance (either overtly or covertly) to all projects eliminates the strategic intent of individual programs and increases overall risk to the organization.

It is also critical that the corporate leaders understand and apply customized expectations to projects depending on maturity level. Projects in the compress phase should receive more grace for missed schedules and budgets than projects in the thrust phase. There are simply more unknown-unknowns early on that need to be worked through.

An organization's risk culture affects management decisions. Notice I did not use the term "risk-averse culture." While it is probably quite clear that I personally prefer risk-seeking culture, having gravitated to the early-phase R&D-oriented projects my whole life, there are no sound data to support that this type of culture has a corner on success. In fact, one could, and some books do, make

"Frankly sir, we're tired of being on the cutting edge of technology."

Figure 14-1 Cutting edge cartoon. Courtesy Randy Glasbergen.

a solid argument that risk-neutral and risk-averse organizations can and do flourish. I admire the Randy Glasbergen cartoon, shown in Figure 14-1.

Risk-ignorant cultures do not understand the nature of risk, why it is healthy, why it is dangerous, where it fits into the business model, and where it should be avoided. Risk-ignorant cultures confuse various risk signals, including the following:

- Risky idea versus risky market
- Risky technology versus risky business model
- Risky personality versus innovative personality
- Innovation versus risk
- Correct risks versus inappropriate risks
- Risks that even in failure will lead to success versus risks that in either failure or success will lead to organizational damage or collapse

Management needs to understand the risk–return universe, as shown earlier in Figure 8-3. Perhaps a more literal view of the risk–return universe will help explain the natural, healthy tension between the risk side and the return side. Figure 14-2 depicts this tension by showing the types of risk typically represented by a risk-managed environment and an exponentially innovative environment. These are the forces the typical project manager has to deal with to effect a logical outcome.

Risk-Managed Development
- Linear
- Controlled
- Predictable
- Measured
- Guaranteed
- Incremental
- Less risky!
  Or maybe not.

Exponential Innovative Development
- Nonlinear
- Uncontrolled
- Unpredictable
- Difficult to measure
- High % failure
- Discontinuous
- High risk *or* low risk (opportunity cost of not taking risk)!

You need both to succeed as a company, employee, product, customer, industry!

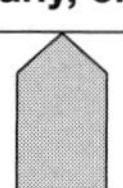

Figure 14-2 Tensions between risk and return.

The risk–return universe and tensions graphic explains the fundamental tug between culture and policies and the organization's need to grow into new markets with new products and services. The figure explains most of the errors in judgment that exist when people and organizations are faced with new developments, market forces, or technologies. It also helps explain the life cycle concept. For example, DuPont's Teflon was invented in 1954 but was largely underutilized for years. Through a long series of business and technical decisions, it was finally commercially developed by an outside firm in 1986 as SilverStone antistick cooking ware. The Teflon story represents the path from accidental R&D breakthrough to purposed, developed commercial product. Today Teflon is used in hundreds of commercial and industrial products and continues to be developed for new applications.

Traditional risk management, which grades consequence and likelihood, often adds measures to a designer's/researcher's plate that, by definition, are in conflict with the development of new and breakthrough concepts. Many designers and scientists questioned during the SRM Survey reported that they were "measured" and graded on their adherence to program management control boundaries (cost, schedule, risk) but not on their ability to solve complex problems with elegant, innovative solutions. A few designers did not report this phenomenon. When asked directly about it, they said that they felt as though their organization understood the intrinsic nature and importance of managed risk. Two of these organizations are doing very well in their respective industries today. One is a European automobile company that is flourishing while other automakers struggle; another is an international aerospace firm.

I believe that real success, sustained throughout the Innovation Lifecycle, requires more than a basic understanding of the risk–return universe and good opportunistic program management. There are some hidden keys below these basic management tenets. One powerful key for an organization to understand and embrace is "phase." Understanding the relationship between phase, risk, return, and innovation is the Holy Grail of successful organizations.

The understanding of phase is the delta between many successful programs and the rest. Phase, specifically *the organization's understanding of the strategic goals of a particular phase*, is a key ingredient missing in modern risk management, modern strategic planning, and modern program management and in their respective tools. Simply put, there are different risks and different risk levels expected and needed in the intake, compress, combust, and thrust stages, and those risk levels vary widely based on the market/product development strategy of the organization.

To complete 3DRM and Intelligent Innovation, a third dimension is added: program phase and strategic intent. These elements, when added to the traditional risk management elements of likelihood and consequence, help solve the problem. A return or opportunity aspect is built into the phase dimension, as you will see later.

Understanding and developing phase-specific strategies and tactics is only the first step. Developing organizational policies and procedures that match each phase and even hiring different temperaments to manage each phase are more advanced concepts that truly pay dividends.

Typically, the first phase, the intake, focuses on fostering (discontinuous or radical) innovation and long-term strategic goals, while subsequent phases focus more and more on balancing innovation, cost, schedule, and technical performance on the other end. The last phase, thrust, puts priority on execution, well above cost or innovative needs. *In the last phase, process innovation comes before product innovation.* Here any innovative solution to a schedule, delivery, or distribution problem is weighted high above a material or financial choice or ideal. Putting the product into the marketplace is the only way to produce thrust. It is the only thing that matters in this stage (aside from, of course, quality, safety, and ethics concerns).

To highlight the differences between these phases clearly, and to exploit the unique opportunities available in each phase, the firm must examine the risk constituents of each phase. For clarity, risk is defined as a potential future situation that has a unique likelihood and a unique consequence different from the goals of the organization. I do not define risk in more typical terms of "failure."

Almost any event in a project or program has a likelihood of occurrence. Typical events such as "receive funding" or "finalize design" have robust likelihoods (a high percentage probability of occurrence) because of the comprehension of those events and the overall commitment of the firm to achieving those events.

However, most risks are much less understood and are related to individual technologies or design parameters that are either highly volatile or have little or no historical basis from which to estimate the probability of occurrence. These may have a low or high likelihood of occurrence. In either case, the project manager typically gathers the team together. The team puts forth its best estimate of the situation and begins tracking the risk. Since likelihood is a simple probability, there is little need to alter its use or mathematical bias.

Consequence is a totally different component. Rooted in historical and cultural biases, consequence can be but is rarely linked to broader project or organizational goals. Fine-tuning the organization's understanding and use of "consequence" is explained in the rest of this chapter. Likelihood is a more basic component and is good enough as a simple probability.

3DRM takes the traditional likelihood and consequence chart and changes the definition of "consequence" to more appropriately match the program phase and the organization's strategic goals and needs of that phase. 3DRM links levels of risk and the elements that contribute to success in the organization at that phase. This approach is directly compatible with the classic systems engineering dogma of linking requirements to goals and design.

Before continuing, each organization should use the quick self-diagnosis tool shown in Appendix E to determine the organization's risk tolerance and overall understanding of the risk–return universe. Following is an excerpt from the original study questionnaire, containing questions that allow managers to assess the severity of their specific situation. These questions alert managers if their risk program or program management tools are hindering overall performance.

- Is the project budget based on technical results linked to goals or calendar or worker hours expended? *I-Factor consideration*: If your project loses budget on a yearly basis, rather than by project cycle or project goals, there is a high probability this is affecting the ultimate innovative yield.
- Does your organization consider budget in risk calculation? *I-Factor consideration:* If the answer is yes, the practice is probably affecting the overall accuracy of the risk plan.
- Have program metrics such as RM, CM, EVM, and so on been used in the evaluation of personnel? *I-Factor consideration:* If so, it is quite possible people are designing in accordance with the metric numbers and not with optimal, innovative, or strategic goals in mind. This is often a tracer for risk-averse organizations. These type of numerical metrics can reduce visibility of the organization's strategic goals and

encourage mediocre incremental designs (which are safe and preserve position and title).

- Have we missed major paradigm shifts in our field that we later must acquire as a catch-up measure?

If the answer is yes to any of these questions, risk management or corporate project management policies (along with other project management measures) may be detrimental to the organization's innovative quotient (I-Quotient) and overall performance. *Detrimental* is defined as any processes or constraint that hinders strategic intent and prevents attainment of the "elegant solution" unnecessarily or prematurely. Obviously, poor program management of any type is also detrimental, including poor requirements definition, totally inadequate staffing or budget, lack of metrics, and so forth. I will not elaborate further here; business and technical history annals are full of stories on both sides of these issues.

Success is defined as product development efforts (again, product development represents a range of hard and soft goods and services) that have had a profitable and sustainable stay in the marketplace. The marketplace for many solutions may be hidden within other products—business-to-business (B2B), government, and so on—and is not necessarily limited to consumer goods and services. DoD efforts may have slightly different criteria but can in general be defined by all of the above as well.

EXAMPLES OF PERSPECTIVE: WHEN RISK, OPPORTUNITY, INNOVATION, AND ELEGANCE COLLIDE

The SR-71, a long-range, supersonic aircraft, is one of the most successful designs in the history of aviation and, more broadly, in the history of mechanical design. The aircraft was fitted with cameras to take pictures of a situation or other targets, where, for example, a potential enemy was doing something "of interest." Although now decommissioned, it is credited with preventing at least two wars because the quality and timeliness of the information provided were major factors in helping politicians develop the appropriate negotiation strategies. (See Figure 14-3.) It flew over 17,000 sorties and 3,500 mission sorties and had no failures during its operation from 1972 to 1988.

In almost every discipline (mechanical, electrical, materials, thermodynamics, aerodynamics, electromechanical control, production, schedule, cost, and performance), the SR-71 conception and delivery shattered the boundaries of the day. Each of these boundaries would have been listed as a bright red, potentially certain and catastrophic failure on a traditional risk chart. In today's environment

Figure 14-3 SR-71B at takeoff (Courtesy: NASA).

it is doubtful the program would have survived the first milestone or the preliminary design review (PDR) stage. (Milestones and PDRs are common government-mandated program reviews in the defense industry.) And yet the SR-71 was successfully developed.

The SR-71 and similar projects, in both commercial and DoD segments, push the limits of common project management metrics. The metrics are typically designed to keep everything within the specified parameters and to provide predictable results. These results meet predetermined cost, schedule, and technical boundaries. If this predetermination is incorrect, the project will suffer short-term consequences. If the predetermination does not include the future state that a competitor is working toward, the firm will suffer long-term consequences.

Projects for which there are no boundaries, or for which the boundaries are "beyond description," in the current vernacular, are particularly hard to manage with any tool. The NASA Mars Pathfinder exemplifies a modern-day team that used unconventional methods to accomplished its goals. The team had a severely limited schedule (one-third the length of a normal schedule) and limited budget (1/15 the size of normal), coupled with an organizational desire to succeed and minimal yet solid program management. The end result was an astounding success.

Other examples abound, including such venerable commercial successes as the Mazda Miata, the original Ford Taurus, and the original IBM PC, all rated as high risks and nearly canceled in their original form. All of these products were so innovative that they changed their organizations and respective industries.

Case Study: SR-71—Accepting Risk in the Early Phases

Imagine a fuel system designed to leak! What would the media have said about the engineer who, against all common sense and in direct conflict with a "high risk" report, designed the fuel system to leak? The SR-71 flew so fast that the wings heated up to a bright red glow. Special materials and manufacturing methods had to be developed just to keep the wings together. The fuel tanks in the wings could not withstand the heat and thermal cycling stress. No rubber-type material for sealing the tanks in this environment was yet available, so the program was in jeopardy of grinding to a halt because of the fuel tank problem.

A solution was needed fast that kept the program on schedule, met or exceeded performance requirements, and was safe for the pilot and civilian community.

The team sprang into action, emphasizing the term "elegance" as a mantra for success. The team was careful to not overdesign or underdesign a solution.

The solution was to leave the wing tanks unsealed. If the titanium tanks were assembled in a precise fashion, without sealant, they would be able to move and expand while they heated up and simultaneously form an effective seal for the fuel. This amounted to an innovative, high-risk, high-return solution that normally would never be accepted in a traditional aerospace design culture. No long, detailed mitigation plan was developed; no management reserves were set aside. And yet a fuel system designed to leak at ground level but survive the extreme temperature and pressure differentials at high altitude and speed was the elegant and robust solution. It was innovative in a way no one expected. It kept the program on target, and cost, technical, and schedule milestones were met or exceeded.

Many of today's companies focus primarily on financial risk. Technical risk and schedule risk are related, both culturally and mathematically, through the Capital Asset Pricing Model (CAPM). Suppose, for example, a person wishes to invest in an unproven software technology firm consisting of five brilliant but young computer programmers. The risk in this scenario, both mathematically and intuitively, is very high. Our subconscious expectations in this case are for very innovative cutting-edge solutions that are inherently less robust. The money lent to that firm would necessarily be at a much higher return rate, say, 50 percent ROI, than the money lent to a very large firm for a similar venture. We know the payoff with the large company would be diluted and perhaps delayed because of the sheer inertia of the firm, but it would be more certain and less risky. Our expectation is that the product from the large company would be less cutting-edge but perhaps more thoroughly developed.

The SRM Survey showed that most risk management professionals acknowledge the link between social and professional risk aversion and the impact it has on the professionals' use of risk and management tools. Overall, people admitted to a cultural bias against risky projects and a subconsciously more severe ranking of new, uncharted risky products when compared with clear, linear development projects, regardless of the strategic importance to the firm.

The survey also showed that people on risk teams would grade consequence higher as soon as they saw a high likelihood. There exists a certain inability or unwillingness to embrace ambiguity, even though the risk model is based on it.

While this trend obviously affects the intake stage, it also affects every stage of the turbine model and can bring a firm to a grinding halt. For example, whereas Honda Motor Company was taking huge risks to bring its rust-prone little cars with small but powerful and innovative front-wheel-drive engines to the United States in the 1970 and 1980s, Ford had become so risk averse that it did little but polish the chrome and change the color of artificial wood decals during those same years. Honda introduced innovative improvements each year, while working on their quality and delivery in the back room. Ford hunkered down and followed the advice of lawyers and accountants. Years later, with the benefit of hindsight, it is clear that Honda had moved forward while Ford and many companies in the same position in different industries had stagnated. Ford would have realized greater profits had it more accurately assessed the risk-return trade-off and the need for constant innovation.

SOMETIMES FAILURE IS GOOD: THE ADVANCED CONCEPT OF INVERSE RISK AND OPPORTUNITY MANAGEMENT

The SRM Survey results suggest that the basis of rating likelihood and consequence becomes inaccurate when programs with a high probability of failure are evaluated, particularly developmental or new programs. When technologies are in the very early stages, such as pure research, Defense Advanced Research Products Agency (DARPA) projects, IR&D projects, and so on, the probability rating and resulting consequences make even less practical sense. The ratings cannot take into account multiple failures, which may be required as building blocks to a industry-changing technological breakthrough. The U.S. lifting body program of the 1950s and 1960s is good example of this phenomenon. This program provided a solid basis for the current space shuttle via countless failures and a few successes. In fact, depending on the metrics used and the industry studied, it takes between 300 and 3,000 developmental ideas or programs to yield one successful technology insertion. A study by UGS Corp. titled "Powering Innovation with Knowledge" (2005) suggests that 86 percent of new product ideas never make it to market and of those that do, 50 to 70 percent fail. One study from the University of Central Michigan suggests that the pharmaceutical industry has closer to 6,000 failed ideas for every one drug that is brought to market. These development costs total on average nearly $1 billion over 10 years for each new drug. Certainly, in this and all other industries, there is a lot of failure that is definite. And yet we do not know how to manage and exploit failure. We just incur the costs and let it affect our collective organizational psyche for the worse. *Even definite failure can have strategic and technical importance at times.* It is hard for us as project management professionals to look at the consequence of *not* attempting near-certain failure. That concept goes against the very definition of risk management. The U.S. Air Force and Navy (and most other risk management program users) define risk as having two components: "the probability a negative event will occur and the consequence if that event does occur" (as cited by Marshall Robinette, "An Integrated Approach to Risk Management and Risk Assessment," GEIA Symposium, Wright Patterson Air Force Base, September 25, 2000). The key word here is "negative." In some instances, from both a business and technical perspective, the use of the word "negative" to describe risk is dead wrong. Having significant risk is actually a positive, with its degree of "benefit" proportional to the magnitude of "failure." This view does not show up in consequence of failure (Cf) calculations. Business school texts would call this view "opportunity cost." So, to correct some of this mathematical and cultural "error," we define consequence in a way that accents or encourages a specific type of innovation.

In the risk universe, the tension between risk and return is modeled. Corporate culture and policies are the grounding force in any organization. Policies such as accounting and budgeting practices, hiring practices, and quality control practices can either enable or hinder performance. Designers, engineers, and creative personnel are depicted in the middle of the risk universe shown in Figure 8-3. They soar along in the spaceship Strategy, which is the assembly of people, processes, and resources necessary for a project to run. It is influenced by and reacting to the forces of gravity and of the project goals. In the picture, gravity should not be viewed as "bad" and goals as "good"; they are both necessary and enabling, as are policies and procedures and so on. The tension is necessary, as also shown in Figure 14-2, the balance beam. The designer must negotiate between the two worlds every day to obtain the elegant solution.

The consequence factors of 3DRM are shown in Table 14-1. These factors are based on the short-term needs and long-term goals of the organization or project and are edited by phase. In this example, innovation-enhanced consequence factors are used in quadrant 1. Each succeeding quadrant switches to more traditional levels. Still, each point on the curve represents a risk-return ratio linked to the organization's goals and objectives. This alleviates the problem (and desire) to make the equation work the same on all types of projects, developmental or

Table 14-1 Three-Dimensional Risk, Logical Return by Phase

Phase	Business Basis	Logical Return (Consequence)	3DRM Basis	Ex-Ante Benefit—Incalculable Returns
Concept exploration (intake)	Vision driven, strategic	Unknown, long term. Is it a building block? Yes/No.	Opportunity cost can yield inverse risk; known as the "enhanced innovation—inverse or negative risk zone"	If I don't fail, I will fail. Long-term strategies and organizational goals are metrics for consequence. Dollars and traditional success measures do not apply.
Preliminary design (compress)	Mission driven, tactical	Should provide basis for next stage. Measurable contribution to vision or mission.	Use risk to reduce alternatives by corporate strategies; known as the "smart innovation—neutral risk zone"	Mix of consequence metrics linked to long-term strategies and loosened traditional program measures. Begin sifting nonrelevant innovations out, leaving room for high-risk, high-return candidates.
Final or detailed design (combust)	Operational, profit/loss	Definite design solution, must meet projections.	Traditional risk view normally applies, solid measurable performance required; known as the "traditional program management—positive risk zone"	Traditional area of program management measures and metrics. Risk should have a direct, calculable return or be within acceptable levels. Strategic metrics still apply as weighting factors.
Construction and delivery (thrust)	Execution, urgent measured performance	Very high yield, either time or performance; cost not a factor in evaluation.	Smart innovation required, risk tolerance very low; known as the "strict management—enhanced risk zone"	Make-or-break technologies/processes. Innovations are targeted, directly measured, and weighed against delivery. Schedule risk tolerance is low, whereas cost tolerance is high.

mature. This table is just a typical example. Several other iterations can be conceived to realize different results.

ASSUMPTIONS

3DRM is based in part on the following related assumptions:

- Innovation is not the only driver of success.
- Innovation is required throughout the entire product development process. It occurs during the intake (concept exploration) phase as a tool for breaking through current design, performance, material, or method boundaries and through each succeeding phase. During the final thrust stage—construction and assembly—innovation plays an important role in the development of new assembly methods. It is often called upon to counter the effects of incomplete or inaccurate designs that cannot be manufactured as delivered. With the recent rapid reduction in product development cycles, innovation is playing an increasingly important role in these later, nontraditional stages.
- Innovation is usually linked to high risk. However, there are times when innovation, or the adaptation of an innovative technology or solution, is the low-risk alternative. *Conversely, there are times when a very basic, proven, or simple solution is the correct alternative, regardless of risk rating or the desire for a cutting-edge product.*

CONVENTIONS

For simplicity's sake, the typical program phases are presented in Table 14-2 that correspond to various titles in various industries. For example, ISO 9000 suggests a "verification phase," roughly equivalent to our preliminary design "intake and compress" phases. DoD 5000 suggests a "program definition and risk reduction" phase, also roughly equivalent to our compress phase. Use whatever phase definitions make sense for your particular industry or project. The generic 3DRM phases are as shown in the table.

CONVENTIONS: LINKING MITIGATION FACTORS BY PHASE

It is also important to link mitigation factors. The crafting of strategically linked metrics for consequence is directly related to the crafting of strategically linked mitigation plans. The mitigation plans should reflect the same desires, by phase, and actually can be crafted to encourage progress to the next phase if desired. 3DRM includes this as an integral step to the basic process.

Table 14-2 Typical Phases of Program Management

Phase	Typical Conventions
1. Concept exploration (intake)	Idea generation, conceptual drawings, renderings, pure research, simulations, calculations, etc.
2. Preliminary design (intake/compress)	Prototyping, low initial run test, design review, test, simulations, evaluation, redesign
3. Detailed design (compress)	Drawing review and approval, certification, final design, clinicals, one-off manufacture, assembly test, etc.
4. Construction (combust/ thrust)	Final assembly, review and test, delivery, setup, etc.
Post-delivery (thrust) not addressed in this book but can be added to 3DRM	Sale, warranty, consumer feedback, market feedback, recycling, etc.

FIRST TO MARKET: AN EXAMPLE OF TAILORED CONSEQUENCE FACTORS

A Taiwanese electronics manufacturer has a corporate goal of "first to market with lithium ion technology." The program management group decomposes that goal and applies the priority weight to a specific project's risk management program. In this case, "first to market" would dictate a risk-averse, incremental product development strategy based on readily available technologies. By contrast, a "leading-edge" goal would dictate that priority is given to risky technologies rather than technology maturity or schedule. In this case, the goal and its weight may conflict heavily with the director of R&D's desire to develop truly innovative, ground-breaking lithium ion technology. However, corporate goals are clear and would be reflected in the 3DRM metrics. Program metrics are adjusted to accommodate the corporate view, including personnel review policies. This is an intelligent way to influence people, process, and product.

See Appendix B for a fully developed example using all four phases and the full 3DRM method.

APPLICATION AND ORGANIZATIONS

The 3DRM risk enterprise portfolio graph, shown in Figure 14-4, is helpful as a general reference for any project or program manager during any phase of the project. Once fully developed and linked to the organization, each risk item can become a significant factor in fostering innovation and intelligent risk and discouraging unnecessary risk. When linked to the Strategic Balancing method, the

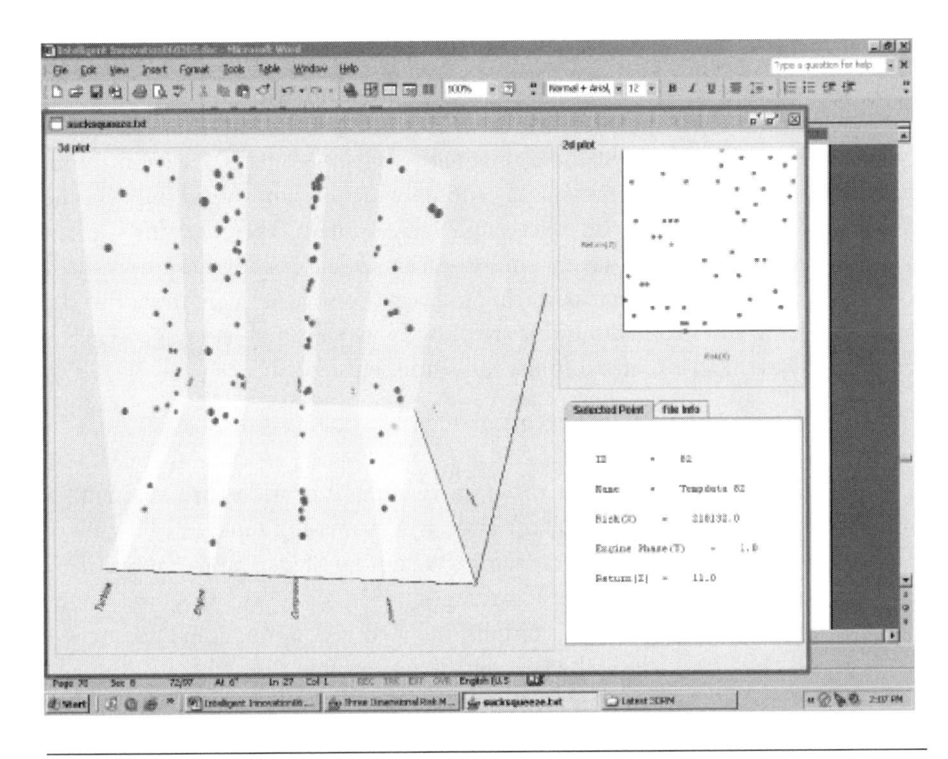

Figure 14-4 3DRM risk enterprise portfolio graph.

four engine stages can be influenced positively to achieve specific goals of the organization's vision and mission.

Another step in advanced portfolio management is postulated in this section. Since we know that market value is related to innovation, innovation is related to risk tolerance, risk tolerance is related to investment, and investment requires a portfolio view, we can also surmise that the return on the portfolio can be managed toward specific goals. It may be possible to increase portfolio yield (and, ultimately, stock price) by investing in a portfolio of high-risk ventures that have low covariance or high variance. That model would suggest a conglomerate with a centralized R&D management system that pools early-stage high-risk resources for disparate industries—a very unlikely and ungainly animal. However, continuing on this possibility, suppose this odd monolith could choose to support a higher percentage of high-risk projects in an effort to boost future performance. Now suppose it chooses to lower covariance not by technology but by phase. It could choose to separate projects by phase and fund only the highest-risk/highest-return projects in each phase. Then it could choose to invest in some with medium risk and medium return (and duration), some with low risk and return, and so on. In total, this approach may over time produce a higher total return than a traditional risk managed, non-phase-specific approach.

The graph shown includes arrows representing weighted strategic intent (magnitude and direction) of the individual risk items. This is a view from an actual client rating. This company has chosen to focus innovation only on the funnel intake stage, leaving the other stages largely intact. The company is in an extremely risk-averse industry and must accommodate very strict government controls. In this type of scenario, during the compressor stage, anything but the most promising and logical solutions are passed into the combustion process.

The process can be reduced to the following steps:

1. Identify and decompose corporate vision and mission. Weight each key factor.
2. Identify and decompose program vision and mission (optional but recommended). Link to corporate vision–mission ranks.
3. Identify program success factors. Weight by Steps 1 and 2 above if desired. Redefine failure accordingly.
4. Identify strategies for accomplishment of vision, mission, and success factors. Include the I-Factors if appropriate.
5. Separate success factors by program and project phase (unique by industry and by program). Use the concept of switch points if desired.
6. Use program success factors and strategies to craft consequence factors (also called logical return). Apply weighting. Link mitigation plan development to program success factors, strategies, and consequence switch points. Craft a 3D risk–logical return chart.
7. Develop risk reference charts detailing likelihood and consequence factors by work breakdown structure element or project element. This step is optional but is good for detailed or multifaceted programs.
8. Use the three-dimensional graph to look at trends, overlaps, and gaps in service/product offerings. If an organization has a preponderance of projects in one phase but few in the other phases, perhaps there is a gap or a pending gap. It is quite easy to predict when projects will mature and move from one phase to the next, and therefore one can predict if a significant gap in income (thrust) or other issues may occur. Likewise, technology areas or other metrics can be substituted for phase and the exercise rerun to see whether the portfolio supports organizational goals, both near and long term.
9. Next, the risk management professional follows the company's normal risk assessment process. This typically comprises the following:

er, we* will provide proof of factor 10.
of financial wrangling, merger mania, or technology hype can
fit of consistent new product introductions based on innovation
and process. In other words, the engine cannot function without

as leaving Amsterdam en route to St. Louis in the United States.
ing about getting out of consulting because I was tired of air-
thinking of my mysterious friend and how she was doing. I had
my manuscript for months and was nearing a close. This time I
on for her. I would deliver it deftly, like a Japanese master chef
good meat from the jowls of the poisonous blowfish.

g into the wireless airport free WAN, I see an e-mail from her
ompany. The e-mail read simply, "So?" She likes to be so mys-
brief. I answered simply "Got it all. You need to buy the book to
t! But one more question this time for you: Why bother?"

swered back immediately: "Value." Value? One word! I ask
aning to life and she sends a single word! I sighed. This is
oying.

on a narrow study to answer this very question: Why bother? Why
isk, expenditure, heartache, and trouble of bring new innovative
vices to the world. What value does all this have? The stock price
n study disclosed a connection between overall corporate per-
roduct development. The study is visible, verifiable, and repeat-
ver 15 years of public data in two industries. The results indicate
ign (which includes the actual design, the quality of the product,
ice, etc.) alone is not sufficient for improved stock price (one
all corporate performance), but that stock price growth does not
nificant (new and innovative) products are being consistently
fielded. There is a subtle and very important distinction in this
lls us that even very good products reach a saturation point, par-
e stock investment community. Their valuation is included in the
company already. So even a great cash cow, a high-performing,

am deeply indebted to Mr. Paul Sabin, a talented engineer, designer, and research
elped provide a substantial portion of the research that forms the basis of this

- Identify
- Assess (Cf * Pf = Rf) (Consequence of failure * Probability failure = Risk factor)
- Mitigate
- Track
- Close

10. Summarize the total risk exposure using the 3D enterprise portfolio graph. Eliminate or combine duplicates, adjust conflicts, and mitigate unreasonable exposures.
11. Continue, refine, and adjust.

Many contemporary risk managers are finding that application of risk management to their organization requires constant monitoring. It requires encouragement of early team involvement, training, and approval and reapprovals by top management. In fact, effective risk management requires that the organizational processes are changed for true integration and success. 3DRM is no different and can be more difficult to apply if the proper groundwork is not applied. Since 3DRM is a function of the organization's strategic plan and goals, it should be inherently integrated with senior management's desires and other processes from the start, which will facilitate its acceptance. 3DRM, like regular risk management, should be promoted as a proactive, positive influence innovative management tool. It is a profit-producing activity, not a cost or chore.

CHAPTER REVIEW

Innovations can be high risk, low risk, and everything in between. They must be managed along with all the other activities and investments going on at the time. However, because of cultural and historical norms within a company, managing almost any innovation often gets bundled with managing high risks, regardless of the actual level of risk or the appropriate expectations for that phase of the development. 3DRM provides a method and, more broadly, a framework for assessing all risks with a phase-appropriate, strategically targeted consequence factor. This method includes the understanding of the need for failure on the way to success (analogous to the bypass in a turbine process) and allows for opportunity costs of not taking risks. Once the consequences for an organization are customized with regard to the vision, mission, and strategy of the organization by phase or maturity level, all risks and innovations can be parsed and managed for optimum performance.

MAKING IT REAL

Follow these steps and refine and reiterate as necessary until your organization has developed the correct set of parameters:

1. Train employees on the basic nature of risks and risk management. Warn them about the human nature tendency to rate all risks high, regardless of actual risk.
2. Have management take a risk tolerance test. Note your organizational risk distribution and conduct discussions to uncover the various risk profiles. Help the organization understand who is risk averse and who is risk seeking so that people will better understand one another's behavior and approach to various projects.
3. Develop specific consequence factors for your organization by phase (whatever phases are appropriate for your industry). Be sure to craft these consequence factors to include your strategic intents for each phase, keeping an eye toward propelling innovations through the Innovation Lifecycle. Try making the consequences slightly less catastrophic at each phase to encourage innovation throughput in non-critical (non-life-threatening) industries, and try doing the opposite in critical industries such as medical and transportation.
4. Commit enough funds and manpower to perform a quarterly or yearly risk portfolio. Rate all of the projects and programs and developments in your pipeline using the specially crafted consequence factors. Put all of this information in the 3DRM graph and analyze the portfolio. Look at your "sweet spot." Is it where you expected? Is it where you desired? Look at the areas where you have few projects under development. Are the number and level of risk what you expected? Are there enough projects at each phase to keep the engine running?

A customized portfolio management mentality and process should emerge. The organization's comfort level with the terms, tools, and theories will take some number of months and iterations to raise to an acceptable level. However, the adjustment is worth the trouble. This method can alert people to blind spots and judgment errors that typically occur when only looking at a project in two dimensions.

STOCKING UP ON INNOVATION

"By 2010, products representing more
will be obsolete due to changing cust
offerings."

—**Deloitte Research LLC Glol**
Innovation," **Exploiting Ideas fo**

So why bother? Why go through all this pair
products and services to market? Why exp
funds, and valuable production capacity for
high risk? Why stress already tired employ
financial support people and bankers to ext
aren't these the ultimate questions? A few of
the thrill and pride associated with forming
entrepreneurial spirit, but still, none of tha
keeps the creditors at bay, makes the shareh
coming back. This chapter discusses the dire
valuation in both the financial market and t

Answering "why bother?" brings us to o

I-Factor 10: Good new product desig
the long term, sustain corporations.

In this ch
No amo
replace the b
in both prod
fuel and air.

This time
Besides t
planes, I
worked c
had a qu
cutting t

Plug
consultin
terious a
hear the

She
her the
getting a

I embark
go through th
products and
versus innova
formance and
able. It covers
that product
performance,
measure of o
occur unless
developed an
conclusion. It
ticularly with
valuation of t

* In this chapte
assistant, who
chapter.

well-liked product, is not going to bump up a stock price; it already did. These established products are part of the current valuation and no amount of face-lifting and fine-tuning will make a significant movement in overall organizational value. It is the new (and high-performing, well-designed, ethical and safe, and exciting) products that increase stock price.

In this chapter we are starting to put it all together. Here we will clearly see that all the trouble is worth it. The company/engine runs with tremendous thrust, day after day, season after season, when it is committed to new development. We see that investment in time, money, and energy when linked to innovation and the pursuit of new products pays off in the marketplace, in the boardroom, and in stock price.

All studies start with a hypothesis. We hypothesized that good new product design, development, and fielding sustain corporations over the long term. If this hypothesis is valid, then the connection between overall corporate performance (as measured by stock performance) and product development can be measured. Of course, there are other ways to measure overall corporate performance, but this way provided a verifiable, repeatable method. It allowed us to simply quantify the broad topic of product development and its relationship to corporate performance in aggregate terms.

The results of our study indicate that product design alone is not sufficient for improving stock performance (one measure of overall corporate performance), *but that stock price growth and improved corporate health will not occur unless new, significant, and successful products are being developed and fielded.*

THE SELECTION PROCESS

To carry out a study such as this, we had to pick a few publicly traded companies to examine. Selecting the right corporations was important. Corporations were chosen based on three criteria. The corporation must have the following:

- A long operating history
- A product-driven market
- Public awareness of products

A long operating history was important so performance could be reviewed over many product introductions and refresh cycles. Since product development cycles in the auto industry can span from one to several years, each company was reviewed over 13 to 15 years, which we believed would be sufficient to allow meaningful relationships to be identifiable. In addition, because corporate performance changes take time to respond to product development work, the long history allows time for actions and reactions to occur. This aspect of the study also represents the totality of the engine model. It shows a repetitive cycle of ideas, inventions, designs, and goals being met, going from intake to thrust. Some firms actually can achieve financial success by exploiting only one or two stages of the engine for a short period of time. The study goes beyond these temporal results and focuses on firms that have shown results, management, process, products, and survival—good or bad—for a long period of time, The firm needed to provide proof that it utilized each stage of the engine in some way.

The selected companies were required to have a product-driven market. They must compete primarily by product design and not primarily by technology research and development. This definition excludes companies that include commodities, raw materials, subsystems and components, and services. Such companies often recoup large initial investments (plant, R&D, patents) with small (in relative percentage margin), not publicly visible returns over many years. To simplify these effects, we chose companies that compete by making direct consumer products instead of prime development companies. For example, Sony would be a better candidate for our study than DuPont, although product introduction is also critical to DuPont. Xerox would fall in the middle, a potential candidate, but with very complex mixed market and product forces.

Finally, the companies' products had to have public awareness. To separate significant products from trivial releases, such as a "product refresh," we used the number of mentions in the mainstream press as a data point, as will be detailed in the next section. To use this as an identifier (and its obvious link to stock price and the perfect market theories), the companies must be of *public concern* for their new products to be covered in the mainstream press.

Although many other measurement methods are available, the press/stock price relationship has many advantages. First, it can be applied across industry and global boundaries. Second, press and stock industry methods have been consistently applied for over 50 years. Finally, the data are scrutinized by third-party auditors and are readily available.

Public vs. Private

We needed public companies to study in order to obtain the data to prove or disprove our thesis. However, private companies are subject to the same forces and theories. Their "stock price" is directly connected to sales or lack of sales, profits, and even valuation once it is time to divest or go public.

Making the Selection

Since automobile manufacturers met all three of our criteria and consistently use the media for new product introductions, we chose Ford and Chrysler as representatives. We were also interested in a counterbalance to the auto industry. The auto industry loosely represents all medium and heavy manufacturing in this brief study. We needed a study participant that was totally different in as many ways as possible. High-tech companies differ from the heavy manufacturing, assembly, outsourcing, two- to five-year development cycles, and advanced development tools of the auto industry. To explore this differentiation, we chose Apple Corporation.

Apple is an interesting high-tech company because it has had varied financial success but has generally been regarded highly as an innovator in product design. Several times in Apple's history it has outinnovated competitors but failed to capitalize financially because of competitors that outmarketed or outdistributed it. In a sense, Apple filled the funnel, funded the compressor, and ignited the combustion, but it did not get the necessary feedback from the thrust to make it all worth the effort. Apple and the computer industry differ from the automotive industry in some key areas; they use light manufacturing and assembly methods with much faster new product cycles. They also rely on outsourced components even more than the auto industry and operate on slimmer profit margins. We hypothesized that the high-tech market's faster development cycles would have sharper responses in the stock market and, therefore, may have clearer relationships between product introductions and stock performance. Apple would provide a good check and balance for our study and hypothesis. Apple's wild stock swings also provide some interesting insight, and ultimate validation of our hypothesis.

The three criteria used, especially the requirement for a long operating history, tended to lead to the selection of larger corporations such as the three ultimately chosen. We expect our conclusions to hold for small and mid-cap companies as well, and we have preliminary data that show an even more sensitive response, although none is presented here.

IDENTIFYING KEY PRODUCTS

Key products for each of the selected corporations were identified by searching library databases for articles about each company from 1985 to 2000. ProQuest Company's powerful search database provided much of the initial data. For Chrysler that data was terminated at the merger with Daimler-Benz in October 1998. The next step involved scanning each article for mentions of new products.

We noted 13 product mentions for Chrysler, 24 mentions for Ford, and 20 mentions for Apple. The products were recorded by the date of the new product announcement (instead of first product shipment), because the announcement indicates when press articles would begin to affect consumer and investor opinions and activities. It is understood that "Wall Street" institutional investors may receive new product news a few days to a few weeks before the general public does. The time differential of a few weeks coincided well enough with our study when the entire life cycle is viewed over a 13- to 15-year period.

To identify significant products, we then searched for each product with the company's name in the ProQuest database. The search criteria included anytime before the product introduction to one year after the product announcement date. We ranked the products by the number of articles found referencing each product. The top 10 products, over the investigation period, were selected and charted for each company.

This research does not attempt to distinguish successful products from unsuccessful products because of the difficulty in providing a level accounting of the data. Complicated and subjective information such as expert reviews, consumer opinion, internal evaluations, and financial accounting would have to be analyzed and reconciled if one desired to distinguish success from failure. As such, qualitative and subjective reviews, such as those found in popular consumer reporting magazines or Web sites, are often contradictory; for every positive review on a product feature, we were able to find a negative review of that same feature. Our assumption in omitting these data points is a sort of hybridizing of the perfect market theory. Consumers ultimately understand that for every opinion there is an equal and opposite opinion. They assimilate this information and record their decisions whether or not to purchase a product, and the purchase results in a profit or loss, which ends up affecting stock price. So ultimately, the market decides success from failure, and this shows up in profits (loss), which in turn show up in stock price.

To offset the lack of quality grading, "significant" products were chosen rather than good or bad products, where significance was directly related to the volume of press, as described previously. This measure supports or refutes our hypothesis more directly because press mentions directly relate to exposure. In turn, this exposure affects sales, which ultimately translate to profits.

MEASURING CORPORATE PERFORMANCE

Stock price (SP) is a clear measure of corporate value. Although SP is confounded by greater macroeconomic changes, it gives a fair estimate of both current and expected corporate value over time. In theory, SP "absorbs" or reflects press news (good and bad) more than any other indicator and provides a convenient, accredited measure that is readily available.

To measure the relationship between product development and stock price, we used a forward-looking view of stock price. Figures 15-1 through 15-3 reflect the annualized growth rate from the current time to 6 months, 12 months, and 24 months. For example, if the stock grows 6 percent from January 1, 1999, to June 1, 1999, then 12 percent will be recorded on January 1, 1999, to chart annualized growth.

A snapshot view of Ford is provided in Figures 15-1 and 15-2. New product introductions are identified by diamonds at the x-axis and a corresponding vertical line. In almost every case, a corresponding increase in stock price is noted within one to five months.

Using Ford as the model, observe the lag effect in Figure 15-3. The lag is defined as the time it takes the general market to react to the news and *believe* it is real (that is, respond by buying/selling stock). In the auto industry and computer industry, it is common to reveal new products a little at a time, months before the actual product is available to the consumer. The majority of the stock price gains seem to occur in this interim period between initial press releases and when the products are actually released to the buying public, and possibly include the first couple of months in the showroom.

The average increase or decrease in stock price after each major model introduction and the average time lag are significant factors to consider. If one were to calculate this same average by industry, a benchmark could be established for measuring and managing corporate development programs. It could be an NPV prediction of stock price reaction to the future return on R&D projects. Our proprietary database of the 11 I-factors also predicts this "reaction," having been tested along with the SP theory in over 70 major corporate studies. Although managing R&D and innovation and looking for a corresponding reward in stock price is not new, the strength of the link between stock price and innovation and corresponding low variability suggests that corporations can benefit from investing in innovative products/processes/services at every phase of the Innovation Lifecycle. The engine must have fuel and air, as well as management of that fuel–air mixture, at each stage, year after year.

EXAMINING THE RESULTS

Two relationships can be observed in the figures. First, there are several areas of significant product development that coincide with high forward-growth regions.

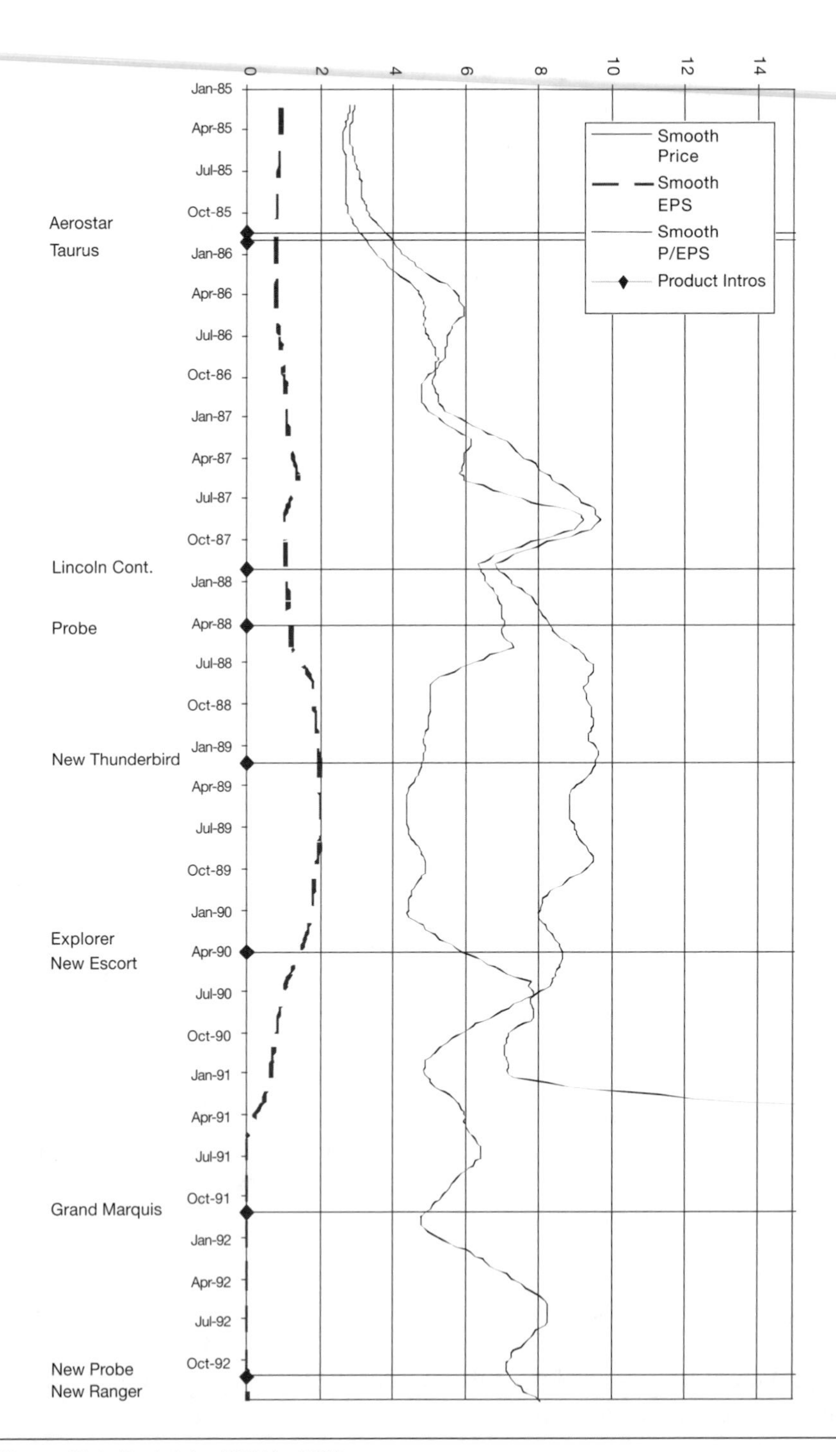

Figure 15-1 Ford data, 1985 to 1992.

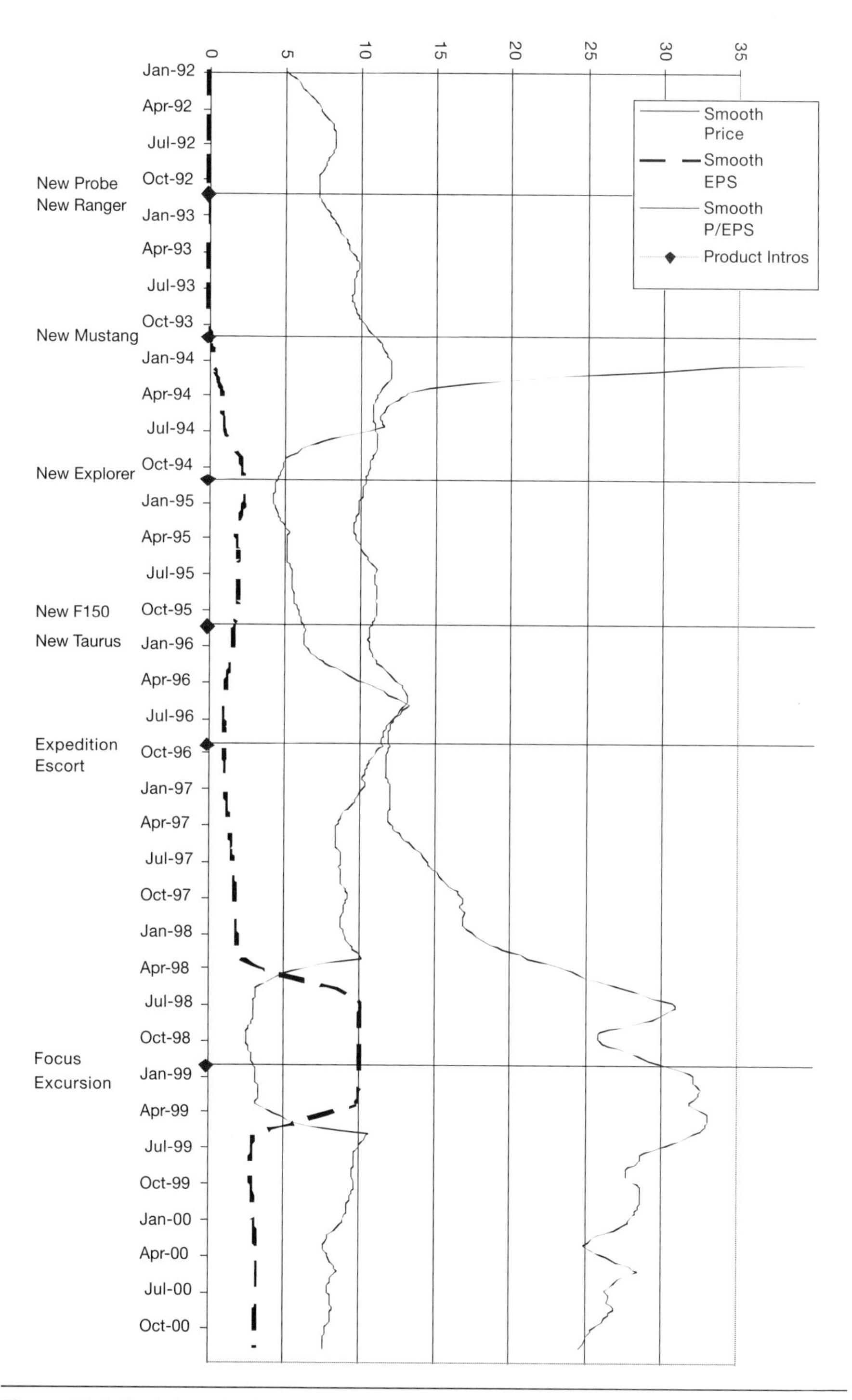

Figure 15-2 Ford data, 1992 to 2000.

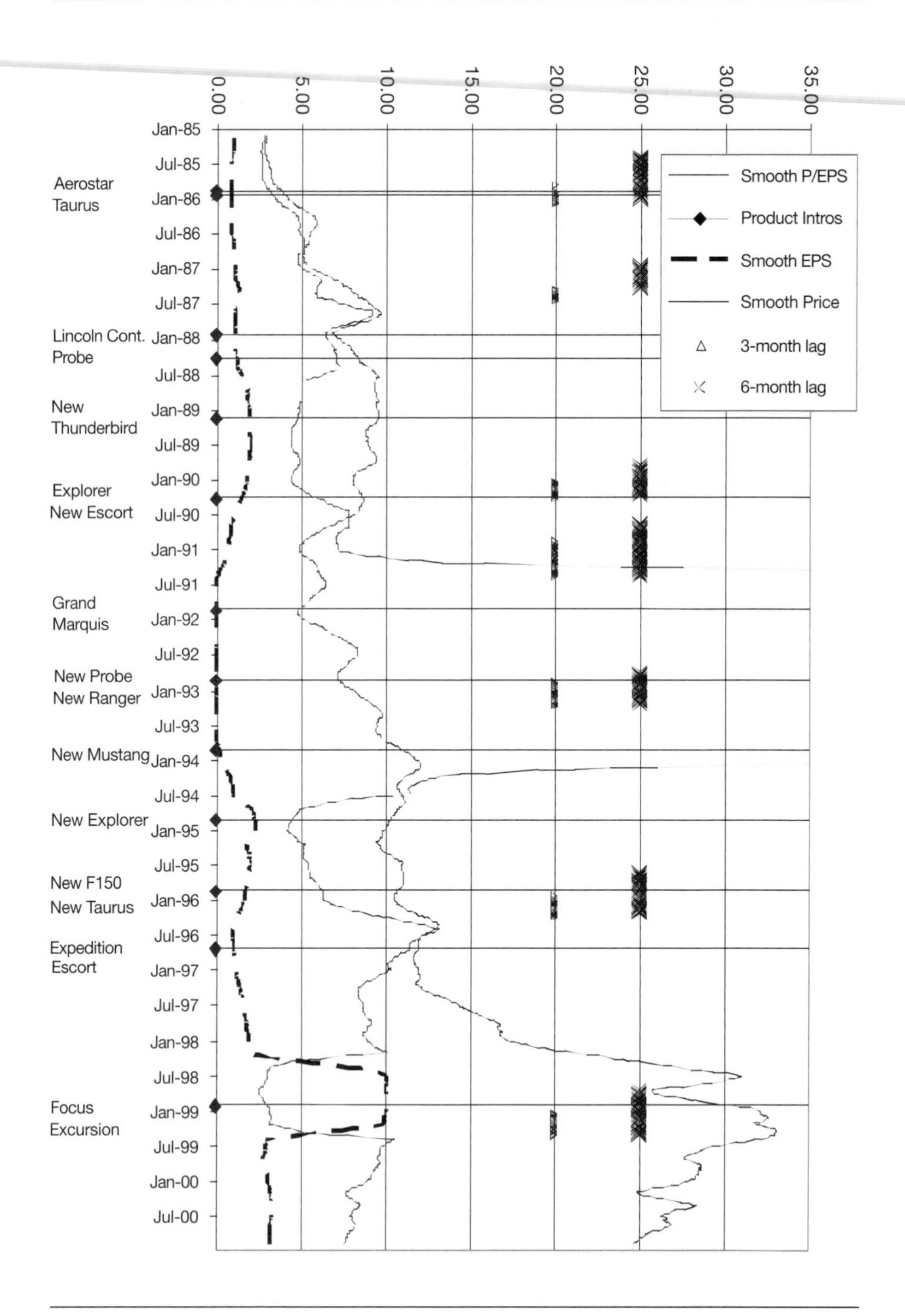

Figure 15-3 Ford data of lagging effects.

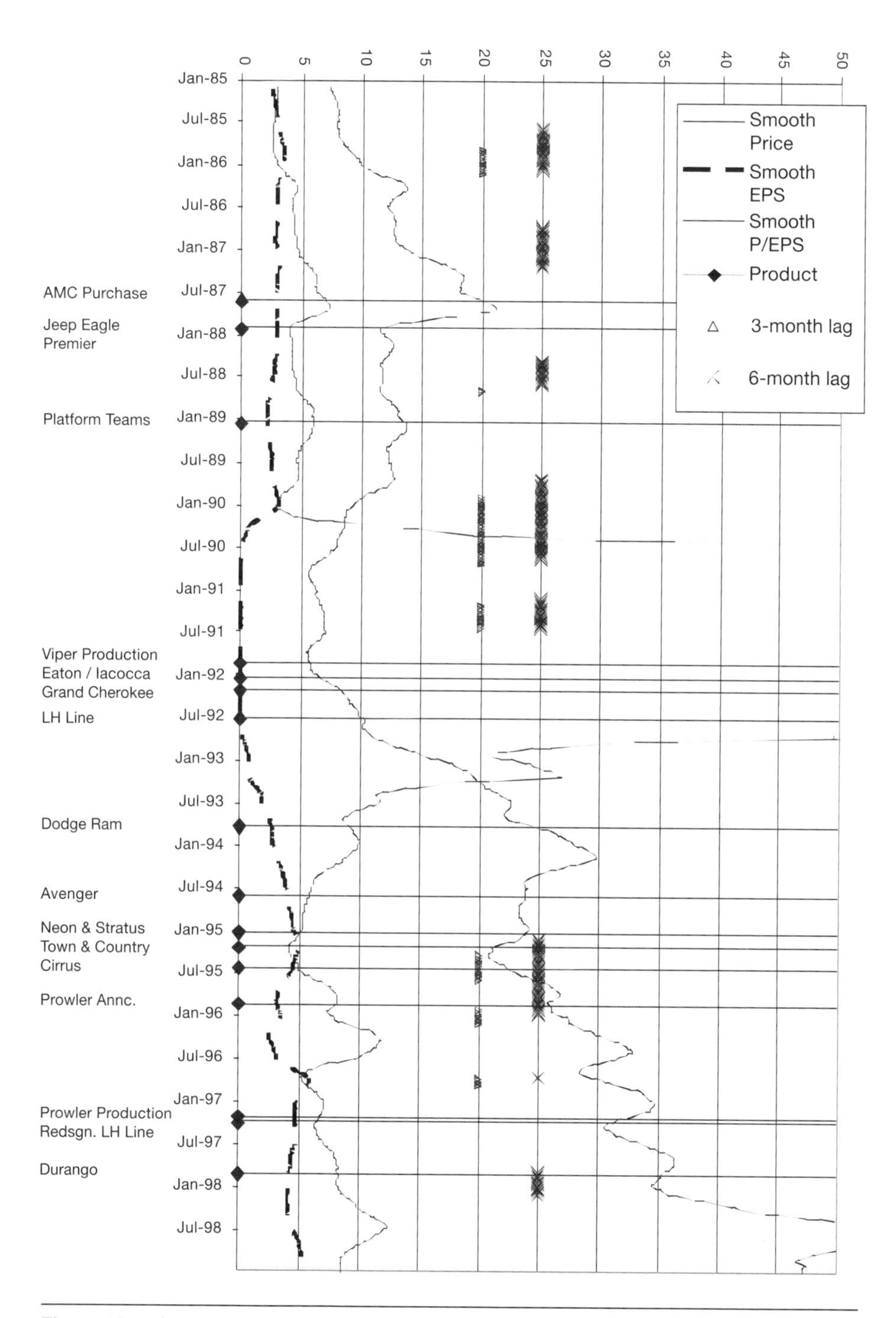

Figure 15-4 Chrysler data, 1985 to 1998.

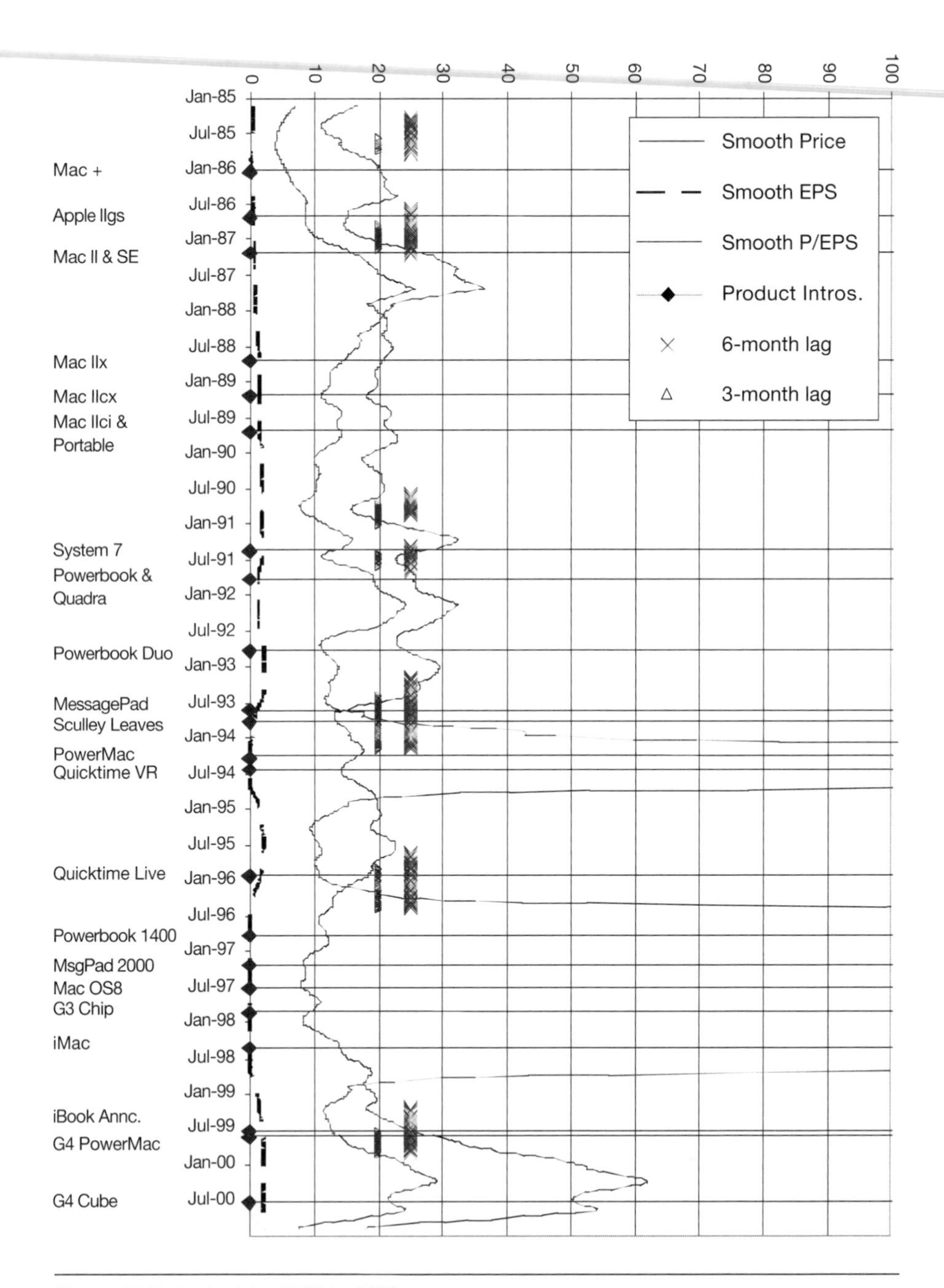

Figure 15-5 Apple data, 1985 to 2000.

This can be seen at Ford in 1986 to 1987, at Chrysler in 1991 to 1992 and less so in 1995, and at Apple in 1997 to 1999 and then in 2005 (see Figures 15-4 and 15-5). Strong product development will lead to growth. A self-reinforcing cycle is at work in the relationship between product development and stock price. Positive elements such as consistent product development, stock price, access to capital, accurate strategies, access to resources of all kinds (tools, talent, process, etc.), and brand development tend to improve the other positive elements.

Each company in our study also had areas of product design with no growth. These areas represent times when products received enough press to make the significant product list but the product may have been (whether real or perceived as) a follow-on product with only marginal improvements. In this case, future earnings are more predictable and may have been already accounted for by investors. Other explanations include that the press coverage of the product release may have been overwhelmingly neutral or negative or that the product may have not been successful in the market because of a major flaw or error in price point or distribution. Additionally, the product may not have been involved in SP changes at all. Perhaps other macroeconomic forces were sweeping the entire industry at that time.

The second relationship seen in the charts is that in areas where significant product design was lacking, forward growth was very slow or did not occur. The relevant periods of stagnation are Ford in 1994 and 2000, Chrysler in 1987 to 1989, and possibly Apple in 1988 to 1989 and 1995 to 1996. *This may be the most relevant relationship for managers.* This is the critical compress and combust section of the engine. If not enough air/fuel—new ideas and products—are let through the intake, the engine will quite literally stall. A stall occurs when the compressor does not have enough air to compress or if the air being compressed is not moving fast enough to reach the proper density. Similarly, the company experiences a stall in the compressor state, blowing out the combustion stage and leaving nothing for the thrust stage, when the intake is starved. The whole engine never produces the anticipated and needed thrust to maintain that self-perpetuating motion.

Another point emerges when trends and averages are viewed. The smaller cap company stocks (Chrysler and Apple) often reacted more quickly to new product development (NPD) news, both positive and negative. One exception (excluding the Apple earnings announcement/computer industry disaster in early 2000) was Ford in early 1990. In this period Ford exhibited a large negative reaction. This period of negative stock growth may be due to the large cash outflow required to launch two models simultaneously, the Explorer refresh and the new Escort, and the subsequent mediocre reception and sales, or it may be due to other wider macroeconomic forces that were hitting the whole auto industry.

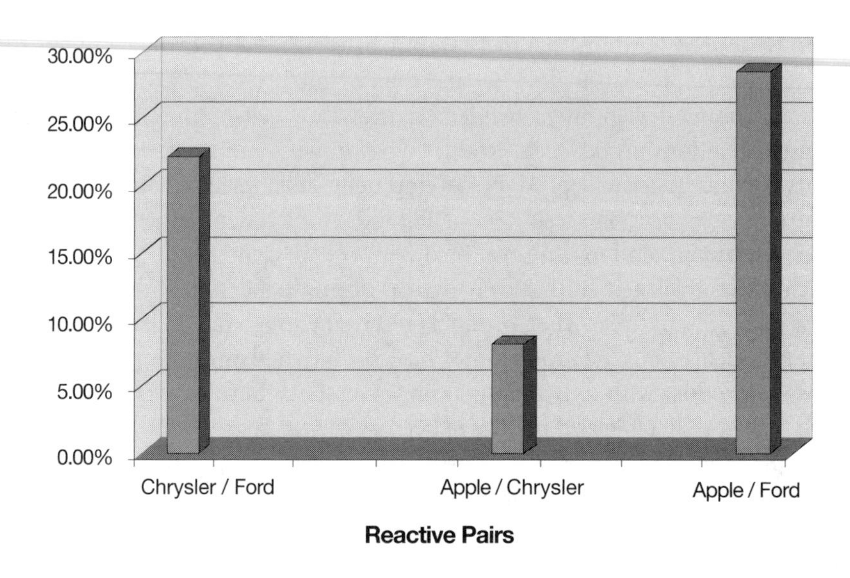

Figure 15-6 Standard deviation as percentage of stock price reaction.

An overall negative trend was occurring in the automotive industry at that time. This study did not separate out larger macroeconomic or industry-specific trends. It was likely a combination of both, using Lexus's growth as a sort of counterweight to the "it's just bad times" argument. Consistent, high-quality NPD can sustain corporations even through the bad times, as was the case with Lexus.

When one views NPD reactivity (the percentage difference in the standard deviation of stock price), a strong correlation emerges among response time, company size, and industry. Apple had the largest standard deviation of $2.66 per share per announcement versus Ford's ratio of $1.90. Almost any high-tech industry leader versus any heavy manufacturer would exhibit similar ratios in response time. In addition, reactiveness can be linked to our 11-point innovation factor scale to produce a force/reaction strategic decision matrix.

In Figure 15-6, we see that Chrysler's movement is 22 percent on average more reactive than Ford's. Apple's movement (less the 2005 iPod-induced jump) is 8 percent on average more reactive than Chrysler's, and Apple is 28 percent on average more reactive than Ford. Imagine setting up corporate NPD and R&D strategies based on this type of data and linking them to long- and short-term plans. That type of strategic planning and management could be a very powerful tool. This is precisely what I started to do with clients in the past year, showing how to graph a firm's I-Quotient and expected SP reactivity linked to R&D funds, risk tolerance, phase, and the like, ultimately showing each phase of the engine and the expected returns.

Errata Sheet: Corrected Pages 161 and 174

The link between risk and NPD is seen in the Apple Computer data in late 1999 and 2000 shown earlier in the chapter in Figure 15-5. A 40-point rise in one year in response to the much-hyped iBook and G4 processor is immediately followed by a punishing 40-point negative market reaction on the news of the much hyped but ill-conceived, poorly executed Cube product—or perhaps it was just bad times for the hypercompetitive computer industry or, most likely, a combination of the two. Although it is difficult to separate broader industry trends from the Cube product itself and the purchase of Steve Jobs' old company, NeXT, it is clear the smaller company reacts more strongly to both types of introductions and investments. The research with Apple continues because the stock price makes such a good visual. Fast-forward to 2005, and Apple shows a dramatic uptick, an increase of over 100 percent in stock price when compared with the 2000 era. Why? What did Apple do when the rest of the computer industry was struggling with saturation, low margins, competition, and commoditization? We will explore this meteoric rise at the end of this chapter.

CONSISTENCY IS THE KEY

The research suggests that, over time, corporations that stopped investing, whether in time, money, or thought, or innovating or managing in any one of these areas eventually saw a significant—if not catastrophic—decline in SP and that consistent, successful new product introduction is a direct contributor to firm valuation, whether real or perceived. A study on Ford from 1985 to 2005 is a case in point. By the 2002–2003 model year, Ford had little to offer compared to its competitors. Its trucks, while holding on to a slim, if not engineered, market lead, were severely dated. GM's trucks had roofs that fold, multiple-position lift gauges, and better mileage. GM also introduced several major new midsized truck models that Ford had no answer to at all. Even Cadillac had the world's most sought-after *truck*. Even now as I write this, with the GM doomsday reporters pumping the TV and radio airwaves full of bad news for GM by the hour, the Escalade continues to sell at full price (and be stolen) at a fantastic rate. Where is the Ford turbine? During this same time frame, Dodge trucks have been consistently upgraded and redesigned, as have Nissan's and Toyota's, although to a lesser extent. Ford's venerable Taurus was so behind its competitors (including the Chevrolet Impala, the Toyota Camry, and the Honda Accord) in both design and value that sales figures were only saved by massive fleet sales (no doubt offered at a bargain).

Taken as a whole, these forces and actions combine to move the stock price, with a significant factor being new product introduction and all it entails. Indeed,

Errata Sheet: Corrected Pages 161 and 174

Table 16-2 I-Factor Self-Evaluator Test Main Question Set

Question Number	Question
1	Personal environment: Innovation totally encouraged (10) or constrained (0) or somewhere in between.
2	Financial environment: Project funding constrained. One-half to two-thirds of original request is deemed optimal. Less than one-half or more than two-thirds and you should downgrade your scores significantly.
3	Time environment: Project schedule constrained. One-half to two-thirds of original request is deemed optimal. Less than one-half or more than two-thirds and you should downgrade your scores significantly.
4	Policies as they relate to Questions 1, 2, and 3.
4.1	Rapid approval of designs and plans: Linked to strategy, timing, and goals of the project, not generic.
4.2	Efficient acquisition of talent, parts, etc.; general consideration, industry dependent.
4.3	Permits "high-risk" or nonconventional approaches to problems without extra review of paper; general consideration co-industry dependent.
5	Use/purchase of commercial off-the-shelf (COTS) technologies.
6	Use/purchase of outside expertise.
7	Risk culture linked to strategic goals; high risks are not overly penalized or praised; risks are seen in relation to corporate strategy.
7.1	Strategic goals affect risk tolerance. Yes: Add 1; No: Subtract 1 to/from score of #7.
7.2	Quick, accessible risk approval process. Yes: Add 1, No: Subtract 1 to/from score of #7.
7.3	Inverse risk understood (the idea of opportunity cost, that some risks and failures are necessary to progress of a technology, art, or business area). Yes: Add 1; No: Subtract 1 to/from score of #7.
8	Clearly defined and integrated corporate vision (where you are going) and mission (what you do).
8.1	Corporate vision and mission match project vision and mission. Yes: Add 1; No: Subtract 1 to/from score of #8.
8.2	Corporate vision and mission are linked to decision making. Yes: Add 1; No: Subtract 1 to/from score of #8.
9	Strategic (how you enact your mission) approach is clear and articulated and matches vision and mission.
10	Step-by-step project planning exists. This is a "5" rating; adjust up or down depending on 10.1 through 10.4.
10.1	Nonlinear projects are OK. Add 1 if OK or subtract 1 if not.
10.2	Parallel processing in project plans is encouraged. Yes: Add 1; No: Subtract 1.
10.3	Projects begin with the end in mind, open to any solution at the beginning with a clear end goal in view. Yes: Add 1; No: Subtract 1.
10.4	Project plans are allowed to be re-based or have baseline flexibility. Yes: Add 1; No: Subtract 1.

The link between risk and NPD is seen in the Apple Computer data in late 1999 and 2000 shown earlier in the chapter in Figure 15-5. A 40-point rise in one year in response to the much-hyped but ill-conceived iBook and G4 processor is immediately followed by a punishing 40-point negative market reaction on the news of the much hyped but ill-conceived, poorly executed Cube product—or perhaps it was just bad times for the hypercompetitive computer industry or, most likely, a combination of the two. Although it is difficult to separate broader industry trends from the Cube product itself and the purchase of Steve Jobs' old company, NeXT, it is clear the smaller company reacts more strongly to both types of introductions and investments. The research with Apple continues because the stock price makes such a good visual. Fast-forward to 2005, and Apple shows a dramatic uptick, an increase of over 100 percent in stock price when compared with the 2000 era. Why? What did Apple do when the rest of the computer industry was struggling with saturation, low margins, competition, and commoditization? We will explore this meteoric rise at the end of this chapter.

CONSISTENCY IS THE KEY

The research suggests that, over time, corporations that stopped investing, whether in time, money, or thought, or innovating or managing in any one of these areas eventually saw a significant—if not catastrophic—decline in SP and that consistent, successful new product introduction is a direct contributor to firm valuation, whether real or perceived. A study on Ford from 1985 to 2005 is a case in point. By the 2002–2003 model year, Ford had little to offer compared to its competitors. Its trucks, while holding on to a slim, if not engineered, market lead, were severely dated. GM's trucks had roofs that fold, multiple-position lift gauges, and better mileage. GM also introduced several major new midsized truck models that Ford had no answer to at all. Even Cadillac had the world's most sought-after *truck*. Even now as I write this, with the GM doomsday reporters pumping the TV and radio airwaves full of bad news for GM by the hour, the Escalade continues to sell at full price (and be stolen) at a fantastic rate. Where is the Ford turbine? During this same time frame, Dodge trucks have been consistently upgraded and redesigned, as have Nissan's and Toyota's, although to a lesser extent. Ford's venerable Taurus was so behind its competitors (including the Chevrolet Impala, the Toyota Camry, and the Honda Accord) in both design and value that sales figures were only saved by massive fleet sales (no doubt offered at a bargain).

Taken as a whole, these forces and actions combine to move the stock price, with a significant factor being new product introduction and all it entails. Indeed, the GM example shows that new product introduction by itself is insufficient for solid stock price. The old adage "do not rest on your laurels" certainly holds true. Ford's exterior design of its new Taurus replacement, the 500, is so conservative, boxy, and mediocre that some publications complained about it offsetting the many innovative safety and convenience features found under the skin. Ford's stock reflects this issue and others, including high part, labor, and retiree costs. These other factors are also a subject of Intelligent Innovation and were explained earlier regarding resource management. To Ford's credit, it is beginning to turn the corner and even use the innovation mantra in its advertisements. Some of Ford's newer models have nice features, like continuously variable transmissions and advanced cross-drilled brakes. While a solid argument can be made that these are not innovations per se, just good features, one could also argue that the delivery of these features at an affordable price to the common market is innovative. Only time and stock price will tell if Ford did the right thing.

INNOVATIONS IN PRODUCT, DELIVERY, AND BUSINESS MODEL

So what did Apple do that broke the mold for the company and stock market, resulting in a 300 percent rise to over $80 per share from the stagnant $20 per share just a few years earlier? What caused a two-for-one stock split in February 2005, the first split in five years and only the third in company history?

Apple introduced the iPod, a sophisticated version of the MP3 player and media storage device. However, Apple's success involves much more than selling a few neat products that were knocked off nearly immediately (although as of January 2006, Apple is still estimated to have between 80 and 90 percent of the MP3 player market, according to a National Public Radio broadcast of January 9, 2006). Apple did something very smart and "discontinuous" within its industry. It went after recurring sales in subscriptions targeting teens, who had increasing amounts of disposable income and were electronically savvy. Apple's popular iTunes legal music download service proved to be a huge success, financially and strategically. Young people want iPods because they want to download tunes from iTunes at 99 cents each. They want iTunes because they have iPods. In addition, adults were attracted to the upscale versions of iPod, and the devices became a popular, near cultlike status symbol. The popularity of the iPod helped sell a few Apple computers too. In the first quarter of 2005, Apple's profit quadrupled, with a whopping 74 percent increase in sales during the 2004 holiday sales rush. During that time Apple sold 4.5 million iPods, accounting for 35 percent of Apple's revenue (and at a hefty margin, probably a similar or higher percentage of its profits). It is continuing to succeed, knocking off its own products, pushing through ideas that were in the intake hopper, expanding features, and selling both "up" and "down" to maintain market share, and Apple deserves many kudos for innovating in a very tough industry. The business model part of the story is perhaps more telling. In February 2006 Apple announced that its popular iTunes service had sold its one billionth song.

It is not known if Apple's management purposely considered a version of the Innovation Lifecycle and planned to garner income in what is termed in this book as the turbine-thrust stage. Apple may have planned to keep the first funnel-intake stage full of products and ideas, as Apple historically has done, and hope for the best in the later stages, with a few hits and a few misses. However, it is clear that Apple was successful and had elements of innovation and follow-through in every stage of the process.

New product design and development (and its corresponding investment in people, process, and perception) is necessary but by itself insufficient for corporate stock price growth. Other conditions affect SP, including risk tolerance, investment choices (particularly IR&D and tools such as CAD, outsourcing, and flexible manufacturing), efficiency and accuracy of the product development efforts and corporation processes, corporate policies, process innovation, exterior market forces, ethics (remember Northrop Grumman, Enron, Ford/Firestone), and so on. Each one of these topics is important enough to warrant an entire study and explanation on its own. The data suggest that corporations must focus on innovative product development to experience stock growth. It follows that strong product development can be self-reinforcing when aligned with other positive corporate elements. The data show that mediocre, poorly targeted, ho-hum new product introductions have little or no effect on firm valuation, both real and perceived. "Successful" new product introductions are innovative, ethical, profitable, high quality, delivered on time, as promised, marketed well, packaged well, and so on. The market demands and deserves excellence. When excellence is delivered, the firm is rewarded in many ways.

This is thrust, plain and simple. Each stage of the engine is working at peak output, propelling the products to the next stage at the correct level of completion, at the right time, in the right ratio. It is a symphony, with no solos.

16

INNOVATION IN CULTURE AND ATTITUDE: REVVING THE ENGINE

"Culture is our set of subconscious assumptions, an organization's collective 'state of mind.'"

—***Adapt or Die: The Imperative for a Culture of Innovation in the United States Army*** *by Brigadier General David A. Fastabend and Robert H. Simpson*

"Every organization has a culture. You either create it or you wake up one day and have it. The practice(s) become the culture."

—***Mark Carneal****, Innovative Resources Consultant Group Inc., 2006*

Intelligent Innovation is an attitude. It pervades every aspect of the firm or organization. Its precepts can be applied to nonprofits, hospitals, schools, churches, and families just as well or better than they can be applied to a classic for-profit, product-oriented firm.

This final chapter leaves you with a quick reference to the forces that affect innovation and development in your firm, along with a few methods to foster the attitude of innovation that are needed to apply the tools. The topics of the earlier chapters—integrated plans, supportive policies, creative thinking, and relentless pursuit of new and existing products—will lead to improved performance and sustained success, but they must be applied by people. Every single aspect of the organization is responsible to contribute toward the goal of Intelligent

Innovation. *In the end, Intelligent Innovation is an attitude that is backed up by processes, tools, and people who work in synergy with that attitude for continued mutual success.*

A self-evaluation tool is provided to assist the manager in evaluating where the organization lies in the world of Intelligent Innovation. The I-Quotient Self-Evaluator™ tool, described in this chapter, is useful, but more important than the actual tool is the approach and self-evaluative thinking that it provokes. This chapter reviews the following:

- What is the first area to attack?
- How do the inverse risk theory and opportunity management affect the organization?
- How are the results of the I-Quotient Self-Evaluator used to improve performance?
- What is the role of culture?
- How does one assess and foster an "attitude of innovation"?

First let me say that in my consulting I have seen the lowest-hanging fruit to be in the area of removing impediments to innovation, impediments to thinking, impediments to forward motion, and impediments to creativity and basic workplace joy. These impediments are often found in the policies and procedures areas of an organization and ultimately affect attitudes and performance.

Policies and procedures that hinder fundamental productivity and employee attitude include oppressive vacation day requirements, ultrastrict sick day or personal day policies, lack of well-rounded feedback in management review sessions, and other cultural and procedural anomalies. They can be in the area of management culture, approval cycles, and even organizational structure—structures that, by their very design, promote unhealthy internal competition that prevents resource sharing and idea flow. This is most common in large companies that are formed by assembling once separate divisions. Management grading and even bonuses often depend on antiquated measures that pit worker against worker. W. L. Gore and Associates—the highly successful makers of GORE-TEX and other related industrial products—is a sort of poster child for corporate governance, culture, and management practices. Numerous articles have been written about its "Rule of 150," that is, how it relies on the peer pressure and efficient communication that is found in groups of 150 employees or fewer. Hidden just below the surface of these discussions is the efficiency of policy that a smaller group affords. My guess is that the efficient, employee- and innovation-friendly policies and procedures at Gore are the egg and the culture is the chicken. More interesting, however, is the implied customization of culture that occurs in each of the Gore plants. Since each plant is relatively small, the culture of a plant can, over the long run, stabilize itself into whatever is most agreeable for the employees of that plant. While I have no specific data to prove this and admit it is a sort of Darwinian assumption, I have not seen anything in the papers written on Gore that would discredit the assumption either. The main point is that culture is not only linked to policy and process, but it can also be nurtured by direct, conscious physical designs such as the type of buildings, the number of employees, executive compensation, management structure, bonus structures, rewards for IP, external partnerships, and so on.

In the areas of research and development, partnerships, and the like, I often find that accounting policies are at the root of failed efforts. Simple mismatches between funding cycles and development cycles can kill motivation and projects alike. Accounting policies and procedures are deeply rooted and perhaps the most difficult to adjust or optimize.

Several of the handwritten comments in the SRM study either directly or indirectly stated this phenomenon. Many of the respondents suggested there was a mismatch between R&D funding cycles and product cycles. Having this as a tip-off to a potentially broader problem was a boon for the research that occurred in the ESRM study. One engineer stated quite simply, "I had a patent that needed follow-up and more research [to determine the extent of its tremendous market value, potential licensees, and partnerships] but could not even get the funding needed to talk to the attorneys." The R&D monies had dried up, with the engineer pushing hard just to squeeze the patent out before exhausting all reasonable measures of his (and his team's) personal clock hours. So when it came time to take a step beyond the R&D phase into some sort of preliminary design and marketing phase, there were no resources. The firm's accounting policies did not allow for continuations; they did not allow for the potential of life after birth. The stages of the engine were not connected.

WHY POLICIES AND PROCEDURES HAVE THE GREATEST PERCENTAGE OF NEGATIVE EFFECT ON INNOVATION ENACTMENT

Policies and procedures that are not friendly to innovation affect overt, covert, intentional, and unintentional innovation efforts equally. Worst of all, they impact employees negatively because they are seen, quite correctly, as immovable. Notice I said "employees" are affected, not the actual innovation or process or product. Policies affect innovation at its root, at the very place it starts and all the way through enactment, because they control behavior of people first.

In short:

> *Good policies promote and bad policies demote people—or at least their attitude and motivation. They directly affect ideological ingenuity one way or the other.*

Whether a policy or idea is good or bad is related to its impact on throughput. Ask yourself, "does the policy/idea hinder or help the engine to breathe?" Keep in mind that rejection of bad ideas is a good thing, so "hindering" does not always have a negative connotation. This is a powerful concept from which to analyze and adjust the status quo. Policies and procedures are beyond the boundaries of a single person, for example a good or a bad boss, and are related to structural elements of the firm. These are often beyond the access of most employees, even mid- to senior-level managers. A bad policy (real or perceived) is a source of frustration and creates a sense of helplessness, nearly instantly dousing any innovative spark that may arise. One of the main squelchers to innovation is a set of rules that maintain the status quo.

Mid- to senior-level managers possess the power to change these issues, but it is not an easy endeavor. A separate book could be written on the psychology of what makes an effective mid- to senior-level manager, what type of person gets to those levels, and why some are more effective than others at changing ineffective policies and procedures. For some, whom we will call group A, change efforts are possible. People in this group got to where they are through hard work, understanding the system, and working with and around it over the years. They are changers, movers, and shakers and yet loyal employees. They tend to want to do the right thing for both the organization and themselves. They believe in their teammates and want rewards for them and will work the system to give them rewards. They are often the first to bump into bad policies or procedures and are also the first to be told "No, that's the way it is." At that point they have a choice: fight or comply. Here again, we see behavior that is split. Some will comply with issue C because they know they want to fight issue D next week and D is more important. This group has a chance of success at change when they band together on a particular issue. This phenomenon is rare—we all deal with subconscious memories of the "mutiny on the Bounty." We all believe mutiny is bad; however, if presented correctly, the mutinous act can be seen as a good thing.

The second group, B, will do nothing. This group of mid- to senior-level managers got to where they are by working the system solely for their own benefit. They can only work inside the box. This group is split between those who actually worked the system, finding loopholes in the policies and procedures that would advance their careers or goals, and those who worked the system by finding ways to hide from it or within it. Either behavior creates a person who will not try to change a policy. These people have an interest in preserving the status quo and often have their power wrapped up in understanding of those policies and procedures, over and above skill, accomplishments, or innovative thinking. They are not thought leaders. Group B people are often a hindrance to innovative success themselves, regardless of any policy or procedure. However, interestingly, they can be, in the right positions and situations, high-performance people. If they are involved on a project or work in a division that is designed to turn the crank repeatedly, they can actually do good. Certain work on cash cow products requires a mentality of rigor and repetition and adherence to existing procedures. If an organization has the luxury to target and move people based on personality and skill types, this would be an advanced Intelligent Innovation activity, splitting up your group A and group B people by task. Unfortunately, most organizations cannot afford to just move people to areas for which they are optimally suited based on personality.

Eliminating bad policies or at least improving the organization's policies, processes, and culture can go a long way toward improving performance. One major U.S. consumer products firm attempted to do just that. In late 2000 the CEO decided to conduct an experiment by forming a separate group within the firm with the specific charter to incubate innovative ideas, policies, processes, technologies, and cultures. This broad and difficult charter was given to a group of 60 people, led by a relatively new employee, himself a former vice president of supply chain management for a well-known global electronics firm. The group was able to go around almost any policy or procedure that was in the way of a project they were working on. Since they worked on both leading-edge core technologies and other technologies that were from other markets, they were involved in nearly every area imaginable. The team worked on actual technologies as well as business methods, software architectures, process analysis tools (to improve the way the industry did self-analysis), and so on. After two years in existence, the group had a few tangible successes. The few tangible successes did not amount to millions or billions of dollars in sales, and yet the experiment was considered overall a whopping success. Part of the success was due to the showcase nature of the group. The CEO used them as a shining example of the skunk works of his firm and visitors loved it. Part of the success was related to the hidden objective of their charter. The CEO really wanted to see what stopped them. He wanted to know what and where innovation was hindered in his company, if at all. After all the rhetoric was stripped away, the bottom line was that this powerful, protected, talented group with a real budget and carte blanche approval was stopped dead in its tracks time and again by three things:

1. A risk-averse corporate culture and type B managers.
2. Policies and procedures that were linked to, supported by, and in support of the culture found in 1.
3. People who attacked foreign ideas and concepts prior to any reasonable review or tangible form. These people view each new idea for its impossibilities rather than its possibilities. I call this impossibility thinking. It is the inverse of ideological ingenuity and intelligent innovation thinking.

To the group's credit, they had found ways around many of these obstacles, but in every case this ingenuity took time and effort, added cost, and reduced the impact of the original idea/method/invention. In some cases, these work-arounds built enduring relationships with others outside the group—secret comrades who believed in the cause and wanted to help. These people knew it was the "right thing to do" and that ultimately it would lead to a more nimble, profitable corporation. In other cases, these work-arounds lead to bitter turf battles, work stoppages, and enemies. In short, group A and group B were both at work. Interestingly, there were an equal number of senior executives in both group A and group B. The CEO, the vice president of contracts, the director of R&D, and the vice president of supply chain management were in group A. The vice president of engineering, the vice president of business development, and others were in group B.

Alliances were sometimes visible and sometimes invisible—that is, when the "Peter principle" took over. In this case, I am not referring to Lawrence J. Peter's principle of the same name that posits that a person will be promoted to the level of his or her incompetence. Instead, I am referring to Peter the apostle of Jesus of Nazareth. Peter was aligned with Jesus and swore loyalty to him for months before Jesus was arrested as a heretic. After the public arrest of Jesus, Peter publicly denied even knowing him three times within a 24-hour period for fear of reprisal. While Peter later went on to be the foundation of the early Christian church, learning and growing from his shame, most of these managers did not. Their allegiances to this group/experiment were directly and temporarily proportional to the perception of success at that moment.

The biggest lesson learned for the money and time well spent on this endeavor was that the firm learned I-Factor 11:

I-Factor 11: A culture of innovation cannot exist in a policy-driven atmosphere of managed risk and artificial (or easily manipulated) accountability.

Entire business mechanisms must be changed, relevant metrics adopted, barriers removed, and performance rewards adjusted for even the most robust innovation efforts to succeed and take root. For world-class performance we must go the extra step, beyond just changing atmospheres and working on policies. We must go into developing an entire organization of intelligent innovators. As discussed, there are several concrete steps one can take, methods one can use, and management and decision-making improvements one can make that all contribute. *But at the end of the day, it is the people that enact all of the systems and policies. People must be innovative and resourceful and desire success.* People must be free to take the leap of faith into the "art of the possible," as I have heard it called. This leads us to this concept called "attitude." Once the impediments are removed, how do we actually promote it? I would venture to say that embracing failure is one of the ways to improve attitude. The nucleus to this is described in Chapter 14 on the three-dimensional risk method. For instance, say an employee's idea is well founded but fails. As a manager, we must pick that person up, dust her off, and say "Atta girl, why not try it again with changes based on what you learned?" Notice that I am talking about people again, not projects. The project failed, but the person did not. The project cannot have an attitude, good or bad, but the person can. Ultimately this gets to the idea of embracing risk, by phase, in the appropriate amounts.

OK, so you now may be thinking, "Yeah, right. I'm going to have a world-class firm, beat all of Wall Street's expectations, and make a profit with a bunch of happy failures. Not a chance." I agree, not a chance. *Obviously there needs to be a preponderance of success to outweigh the (developmental) investment in failures.* Notice I did not say "many successes." One or two blockbuster successes, such as 3M's Post-it Notes or its Scotch-Weld line of epoxies, can pay for a lot of failures. It is a portfolio view based on the bottom line and an organization's short-, medium-, and long-term goals.

Another part of the attitude portfolio is clearly disseminated vision and mission, as discussed in Chapters 7 and 8. Another part is decision delegation to one degree of freedom, as discussed earlier. And so on.

The point is that one can manage and design the organization to have a positive attitude that fosters policies, processes, and culture to the right degree for the firm and industry. This holds true for those who are optimists, pragmatists, realists, and so on. This type of organizational crafting focuses the efforts first on the thing (the policy or procedure, the idea, the product, the market, the business construct), not on the people or the numbers.

SELF-ASSESSMENT HELPS PROVIDE A BASELINE FROM WHICH TO PLAN TO ACT

One way of assessing where you are and where you need to go is presented in this section. The I-Quotient Self-Evaluator has been used hundreds of times to help everyday organizations change and succeed. The tool helps with visualization of information and will assist the user in making critical strategic-level business decisions and is also useful for making tactical and operational level resource allocation decisions.

The I-Quotient Self-Evaluator tool provides a comparison to the "norm" for the individual firm's self-assessment, by project or by portfolio. Utilizing the graph's information to make decisions and possibly alter behaviors, processes, or policies toward various goals (improving innovation efficiency, new product development, firm reputation, etc.) is the final Intelligent Innovation method presented.

This tool and method provide the user with several unique benefits. Taken together, they form a modeling capability for the strategic planner, corporate financier, or resource planner. That modeling capability allows the user to perform what-if and scenario-based analysis, with a direct tie to the behavioral and process-based methods in use at the firm. This link to the behavioral norms, policies, and process methods of the firm is unique to this tool and forms its most powerful benefit.

We have also used the tool to analyze projected stock price fluctuations resulting from various initiatives, both management and product oriented. This tool links all the aspects we have discussed as critical to innovation. Our proprietary database provides the relative rank needed to see where you are and where you need to go. *Ultimately the I-Quotient is a subjective grade as to a company's ability to promote and capitalize on innovation—in both product and process.*

The I-Quotient questionnaires were tabulated and put into a database, which forms the baseline graph. More refined segment data, by type of respondent, industry, and organization size, are also available through the author.

PREPARING TO TAKE THE I-QUOTIENT TEST

Deciding to take the I-Quotient test is the first step. It must be supported by upper and middle management, and the results are most useful if made totally public to the firm but not to the outside world. This is hard to do and risky, so some management structures opt to keep things confidential or to synthesize or summarize results for the populace, which is understandable.

Table 16-1 I-Quotient Self-Evaluator Grading Values

Score	Condition
0	Totally noncompliant and never will be
1	Totally noncompliant but could be some day
2	Noncompliant—Organization has a general understanding of relative condition
3	Marginal compliance—Knuckle dragging with little hope of standing upright in your lifetime
4	Somewhat better than marginal compliance
5	Sometimes compliant
6	Compliant, with conditions and/or policies that conflict
7	Compliant, with policies that hinder that can be amended readily
8	Compliant, without conditions
9	Exhaustively compliant, embracing, integrated
10	Proactively compliant—Forward thinking, policy busting, trailblazing, efficient, leading-edge, world-class

Each firm, organization, family, and so on operates with basic structural supports. Similar to those described in Chapter 8 (profit, reinvestment, short-term goals, long-term needs), these areas are found in all organizations.

In the following broad areas, an organization derives its operations and existence and the I-Quotient test can be applied individually to an area of the firm or globally to the entire firm. The three foundational areas to analyze are resources, supply, and planning. The user must rate each area from 1 to 10 using the scores in Table 16-1.

After the user answers the questions in the Self-Evaluator tool, the post-test analysis step is performed. In this step the user compares several pair-wise instances to obtain a sort of picture of their overall condition.

Some condition pairs to look at include the following:

1. "Expected" result to actual results
2. Future desired result to actual result (creates the demand or pull by projecting current gaps)
3. Projected (if current plans are enacted) result to actual result
4. Competitor or industry expectations to own result of Self-Evaluator tool

Table 16-2 I-Factor Self-Evaluator Test Main Question Set

Question Number	Question
1	Personal environment: Innovation totally encouraged (10) or constrained (0) or somewhere in between.
2	Financial environment: Project funding constrained. One-half to two-thirds of original request is deemed optimal. Less than one-half or more than two-thirds and you should downgrade your scores significantly.
3	Time environment: Project schedule constrained. One-half to two-thirds of original request is deemed optimal. Less than one-half or more than two-thirds and you should downgrade your scores significantly.
4	Policies as they relate to Questions 1, 2, and 3.
4.1	Rapid approval of designs and plans: Linked to strategy, timing, and goals of the project, not generic.
4.2	Efficient acquisition of talent, parts, etc.; general consideration, industry dependent.
4.3	Permits "high-risk" or nonconventional approaches to problems without extra review of paper; general consideration co-industry dependent.
5	Use/purchase of commercial off-the-shelf (COTS) technologies.
6	Use/purchase of outside expertise.
7	Risk culture linked to strategic goals; high risks are not overly penalized or praised; risks are seen in relation to corporate strategy.
7.1	Strategic goals affect risk tolerance. Yes: Add 1; No: Subtract 1 to/from score of #7.
7.2	Quick, accessible risk approval process. Yes: Add 1, No: Subtract 1 to/from score of #7.
7.3	Inverse risk understood (the idea of opportunity cost, that some risks and failures are necessary to progress of a technology, art, or business area). Yes: Add 1; No: Subtract 1 to/from score of #7.
8	Clearly defined and integrated corporate vision (where you are going) and mission (what you do). Yes: Add 1; No: Subtract 1 to/from score of #7.
8.1	Corporate vision and mission match project vision and mission. Yes: Add 1; No: Subtract 1 to/from score of #7.
8.2	Corporate vision and mission are linked to decision making. Yes: Add 1; No: Subtract 1 to/from score of #7.
9	Strategic (how you enact your mission) approach is clear and articulated and matches vision and mission.
10	Step-by-step project planning exists. This is a "5" rating; adjust up or down depending on 10.1 through 10.4.
10.1	Nonlinear projects are OK. Add 1 if OK or subtract 1 if not.
10.2	Parallel processing in project plans is encouraged. Yes: Add 1; No: Subtract 1.
10.3	Projects begin with the end in mind, open to any solution at the beginning with a clear end goal in view. Yes: Add 1; No: Subtract 1.
10.4	Project plans are allowed to be re-based or have baseline flexibility. Yes: Add 1; No: Subtract 1.

Table 16-3 I-Quotient Question Weights

Question Weights	Question
10%	1
15%	2
10%	3
15%	4
5%	5
5%	6
17%	7
8%	8
8%	9
7%	10

Now the fun begins. Gaps are analyzed and compared to current and future plans for improvement. Current and future plans for improvement are compared to funding streams, budgets, manpower availability, equipment availability, and so on. In addition, the Self-Evaluator tool can be run periodically and results compared to expectations. Ultimately, trend analysis on various efforts can be performed to grade the efficacy of the efforts.

Table 16-2 shows the basic Self-Evaluator questions. These questions are those used to form the basis for the I-Quotient Self-Evaluator summary graph. The user organization can edit these questions to better suit the industry or condition of the organization; however, the direct link to the summary data will be lost. The cost/benefit of customizing these questions should be carefully weighed. It is possible that small changes in the questions can be made with little loss of meaning if the editor pays careful attention to the original meaning of the question.

Each of the Self-Evaluator question grades was multiplied by a weight to obtain the final score or "contribution" to the organization's ability or lack of ability to capitalize on innovation. It is crucial to understand that the aggregate final grade(s) in any individual category are applicable to any stage of the engine process. A constraint at any stage will ultimately affect all of the other stages, as the engine is a living organism, and in our analogy, a bad bearing or choked inlet will ultimately affect fuel performance and thrust, even if the problem is temporarily unnoticed.

The weights are provided in Table 16-3. While we consider these weights proprietary data and have asserted our data rights on them in the past, the original, individual buyer of this book is granted a license to see and use these weights. The

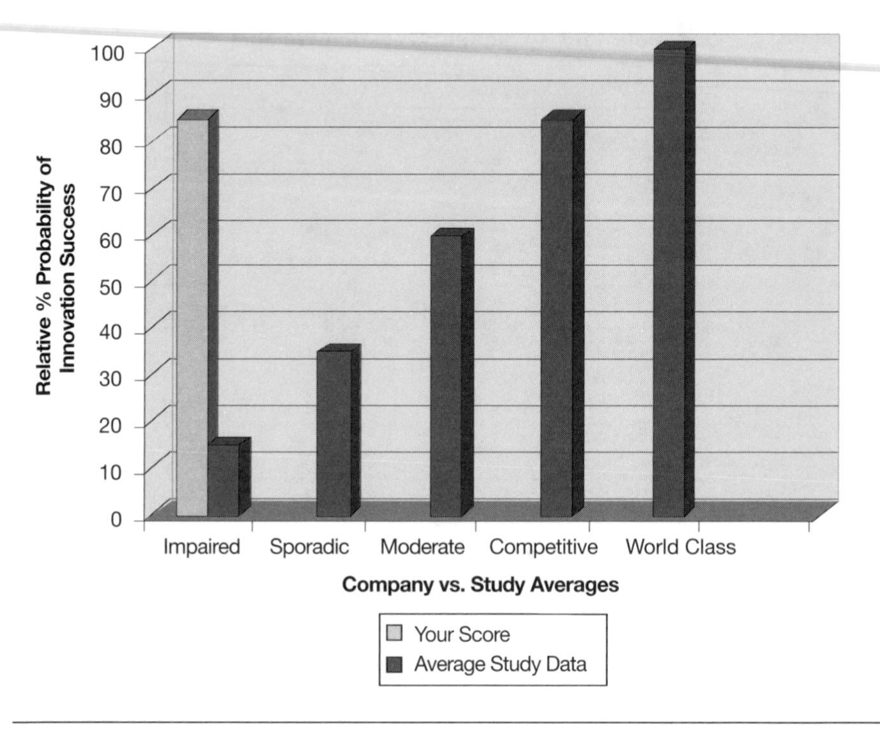

Figure 16-1 I-Quotient Self-Evaluator scores.

user organization can edit these weights to better suit the industry or condition of the organization; however, the direct link to the summary data will be lost. The cost/benefit of customizing these weights should be carefully considered. It is possible that small changes in the weights can be made with little loss of meaning if the editor pays careful attention to the original meaning of the weight. Organizations may obtain a license to use the I-Quotient by contacting the author.

DOING THE MATH

The math behind the I-Quotient is simple. The score for each question is multiplied by the weight, and then they are summed. The power comes in the crafting of the questions and in the comparison to our database of over 1,000 respondents. Figure 16-1 is provided for your use.

At this point, having performed the self-evaluation, the user can compare results to the averages found in the graphic in the figure. This is the overall main graphic resulting from the database. Additional subdivisions by industry or size are available from the author. It may be valuable to stop and take stock in your organization now and try to discern why the self-evaluation score was as high or as low as it is. What is the sensitive point? What one, two, or three issues are most in need of improvement? Which one, two, or three issues are world-class practices to be celebrated and protected? Just having these two groups—at the bullet level—written down on a piece of paper is a powerful tool to effect more detailed discussions and for the development of change plans.

Appendix E provides for a more detailed method of performing the I-Quotient Self-Evaluation on specific areas within the organization. With the original evaluation and these additional evaluations, one can begin to refine improvement strategies and provide a plan that attacks bottlenecks area by area, over a period of months or years.

Keep in mind the principles of systems engineering: The entire system must work together, and super-refining one area without attending to other supporting areas may be counterproductive. Always keep the system in mind. The system is more than the organization; it includes the suppliers and customers, banking systems, and any other entity that affects or is affected by the organization or its products—particularly the people operating them.

Refining or designing any improvement strategy requires knowledge of the environment. The organization can operate without it; however, it will be less efficient without this knowledge. In much the same way, the turbine engine has sensors to determine the density and humidity and temperature of the outside air to refine its exact fuel-air ratio for maximum thrust and minimum fuel burn. Even small differences in outside air temperature are processed by the engine control unit of the turbine to account for the difference in volumetric efficiency. Likewise, the organization must know its customer base, the competition, the global economic forces, government regulations, and the like. Any organization has inputs and outputs, each input has a source, and each output has a destination. This end-to-end thinking is critical in any analysis and planning pursuit. Using this global, holistic thinking, the results and learning that come from the I-Quotient evaluation can be absorbed and acted on. Taken together, these elements provide staying power—the fortitude to move ahead (organizationally, culturally, and financially) in good times and bad.

Finally, using a combination of the methods discussed earlier, the organization can begin to refine its approach, as a whole and area by area. From stage to stage, process to process, beginning to end, your organization needs to think with an innovative can-do attitude to succeed and ultimately excel. The WWII Seabees had a creed: "The difficult we do immediately, the impossible takes a little longer." Perhaps we should all adopt this motto. It is the responsibility of every person, and management has the additional responsibility to innovate in business methods and constructs while ensuring all the phases are linked and managed

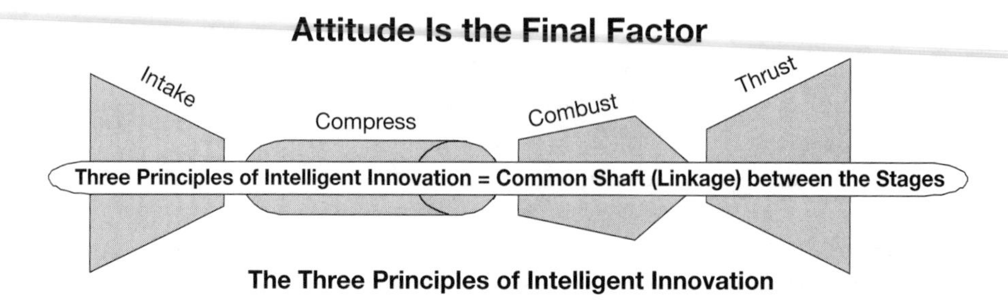

1. Innovation must be pervasive throughout the entire process.
   - It is constant and everywhere in the processes, including policies, tools, and personnel activities.
   - The organization must have a life cycle view of innovation, based on an intuitive sense of the market and a drive to succeed.
   - *Nil Indigne:* Effective decision making, the ability to parse scarce resources (to decide what is worthy), and parallel process are required at each step.
   - Everything is linked to the vision (where you are going) and mission (what you do) of the organization.

2. Innovation is necessary for value growth.
   - Begin with the end in mind: Consistent new product introductions are required for value (stock) growth.
   - Balance risk and return on a strategic fulcrum where strategy is the "how" of the vision.
   - Process innovation, including business construct innovations, is the unsung hero of most blockbuster successes.
   - It is a can-do attitude of resourcefulness and creativity.

3. Innovation transcends all traditional boundaries of products, process, policy, and people. The most successful projects in history all had these traits in common (the secrets to blockbuster success):
   - Limited (severely in some instances) schedule and budget (offset by a can-do or must-do attitude).
   - Technical requirements beyond the current state of the art (offset by creative "possibility" thinking).
   - Shared ideological desire to succeed amongst the participants (such as wartime, intense competition, glory, survival).
   - Capability-based vs. solution-based development, coupled with phase-specific IP, strategic and tactical actions.

Figure 16-2 Summary of Intelligent Innovation principles.

holistically. When management theories and tools have grown dusty and outdated and when the world economy has become truly integrated, the only differentiator that is left is attitude, in both the approach to problems (or opportunities) and their solution.

A summary of principles of Intelligent Innovation is shown in Figure 16-2.

I sent her a copy of the book with the following inscription: "It all matters, but attitude is the greatest of all" As usual, she got the last, brief word. Several months later I got a postcard from Sydney, Australia. It read "yes" on the back.

I never saw her again.

AFTERWORD: THE AMAZING PANAMANIAN RAILROAD

Dario Benedetti is a little more than your average guy—after all, he helped start a railroad. He seems average enough, watching football on TV, playing with his grandkids, and fixing his house. What is most unusual about Dario is his combined traits of insight and vision. They are counterforces: one bottom up, the other top down in nature. This insight permits Dario, and others like him, to look deep into details, into minds and motives, and on the other hand, into cultural and economic situations and to draw conclusions. In the case of the Panamanian Railroad, Dario and several other key individuals caught the same vision and together moved into action.

Vision is like the steroid of business, making people and organizations jump beyond what is currently possible. It is based loosely on economic trends and socioeconomic forces and connects those waves, not dots, to (in the successful cases) place an organization's offerings where and when they need to be in order to ride the wave. The hockey great Wayne Gretsky once said his success was in part because he "skates to where the puck will be." That is a powerful concept, but even more powerful is when an organization attempts to move the net to where its play will be; here vision is working with insight at their highest plane, interconnected with marketing, strategy, product development, and customer influence. The combination of vision and insight is very powerful. Putting the two together in one person or team or one project, with a little luck and some support from other people who caught the virus, is the stuff that legends are made of.

The Panamanian Railroad exemplifies innovation in action. It pits worldwide socioeconomic and macroeconomic forces against the forces of technology, nature, and business. In this story, a few people harness both of those forces and steer them together for good, eliminating the collision and instead building momentum that can be now described in tons per mile. It clearly represents our analogy of a turbine engine with every stage executed and running.

The Panamanian railway company removes loaded containers from ships on one side of the Panama isthmus, transports them via rail—yes, that old-fashioned workhorse called a train—and reloads them onto other ships on the other side of Panama. The total execution of this transport was revived, refined, and relaunched by a small team of people with the support of the Panamanian government.

During the intake or concept exploration phase, our newly forming engine began to sketch out the process. In Dario's words, "the true source of my land bridge introduction occurred in the early 90's while working in Mexico to modernize the rail intermodal terminals." Dario was the director for international business development for Mi-Jack Products, a crane company, and worked on a container land bridge railroad connecting the Pacific port of Salina Cruz to the Gulf port of Coatzacoalcos. Although both ports and the railroad existed, neither port attracted much container traffic, and still don't. Dario says, "Nevertheless, the idea persisted that if you upgraded the ports and rail connection, container vessels could save money by dumping boxes at one port and have them railed across to the other ocean for connecting onward service." And so the seeds were planted. Dario and his partners began to notice other similar forces at work in other areas.

There were other places considered for a transcontinental crossing, but Teddy Roosevelt pushed for Panama and that crossing was built. Other ports have tried to develop this level of business; however, the economics and traffic levels never materialized to the point where they were successful or self-sustaining. The volume transported from Los Angeles to New York via the Panama Canal was too much to compete with.

Dario continued, through the concept exploration phase to the compressor stage: "After exposure to the other Central American and Mexican land bridge concepts, it became clear to me that the same basic physical land bridge ingredients and potential already existed in Panama. Some key additional ingredients included (a) the Canal and Colon Free Zones already attracted abundant container ship calls; (b) the Panamanian Government was planning to privatize the ports and railroad; (c) it was only 47 miles across, a fraction of the length and coincident investment compared to the other land bridge concepts." Dario then outlined the necessary steps while forming a more detailed vision. Subsequently he enlisted the vision, talent, and resources of two key companies and individuals: Mike Lanigan, president of Mi-Jack Products, and Mike Haverty, president of the Kansas City Southern Railway. They both immediately recognized the potential value and supported the venture. Many others provided critical skills and support of the project, and a team was formed. Dario says it best: "As in any project that involves a new idea without precedence, coming up with the concept is important, but a concept never goes beyond that stage unless there are backers . . . who also believe."

Much like the iPod, the land bridge concept is ripe to capitalize on some factors that exist naturally in the universe. It's always better to ride a wave, perhaps kicking it off with your product introduction, than to try to create one from scratch. That did not make it easy. To some the project was absurd. It would require massive amounts of capital, engineering, government support, customer support, and so on. For every believer, there was a polite blank stare. Dario's team was entering the toughest stage of all, the compressor. Here silly ideas are vetted out; they cannot stand the increasing pressure. Here in the compressor, the good ideas get better—ideas are rounded out, additional supporters with unique skills or resources come alongside, and preliminary ideas/designs/fantasies are vetted and straightened. The bad ideas burn out or are sent out as bypass air. Dario and his partners went around the world, talking to government officials, potential customers, suppliers, lawyers, accountants, and so on. Here the technical and business aspects were being vetted through the compressor, along with the political and personal aspects. The Panamanian government and some Panamanian businesses rallied to the cause, seeing the potential not just in dollars but in national pride and the value of a national strategic asset. The idea was being compressed, its potential in thrust getting greater with each new handshake.

Getting through the compressor into the combustor is difficult. Ideas get squeezed, ignited, super-heated, and thrust right into a rotating blade. The rail project was no exception. Potential investors considered the existing pan-isthmus light rail useful for few people with packages to transport. It was more difficult to see the vision of a double-stack, industrial-grade, world-changing transportation system.

The idea of a "system" was the key to understanding the tremendous value of the project. Dario and his team now included some bright local businesspeople. They included employees of the partner company Kansas City Southern Railway and some key government officials, who started talking about the railway as a system, not just a train. Systems thinking is akin to vision as mission thinking is akin to components.

The key players who also adopted this systems thinking included Dave Starling. Mr. Starling helped with the complex operational view and had experience in several aspects of the entire system, which proved invaluable, from international posting to container ocean carrier operations. Mr. Starling is now the president of the Panama Canal Railway Company. Another long-time railroader, Mike Haverty, saw the potential and helped form the original conveyer concept. The component-only view is often referred to as analytical thinking, where the parts take precedence over the system.

Seeing the entire process, from the ships docking, unloading with the Mi-Jack cranes; computer ID and tracking of the cargo, fast and safe insured transport, reloading, legal and tax structures, and all the other components that went into those containers from beginning to end, helped create the end result. *A systems-type project must be planned and optimized from the top down, seen as a whole, and then built from the bottom up.*

When times were tough, Dario says they kept going because "as intermodal transportation executives, the concept of establishing a high volume, 47 mile Pacific–Atlantic rail container link made fundamental sense." They were able to communicate that vision to those around them until the railroad was built, riding, not creating, a wave. Dario continued, "Virtually any mature Panamanian grew up with fond memories of riding the Panama Railroad. Even if they didn't get the container link concept, they all wanted to see the railroad resurrected from its dilapidated status. The grand history of the Panama Railroad, in fact the world's first transcontinental ride, played importantly in motivating all involved. I think the unique land bridge angle, the railroad's proud history and Panama as a global transportation crossroads helped add cachet to the project. With no precedent of any kind, even the financial backers, especially the International Finance Corporation of the World Bank, became believers as well."

Fortunes are made or lost in the fire of the combust and thrust states. Getting the fuel/air mix right and the intake complete gets the organization to the compressor. In the compressor the gases and designs are further refined and perfected prior to being exploded out the back in the thrust stage. Thrust helps harness just enough to keep the intake working and put the rest into forward motion; then the engine and organization work like a dream. Eighteen months after the intake stage, Dario entered the combust phase. Entering this phase was marked by the securing of a 50-year land lease from the government, the last major hurdle. After this passed, the team could see things happening and momentum really began to build. Two years later the railway was almost complete. Although looking at a full-up functioning railroad unloading and reloading containers is a marvelous sight, the innovations are not immediately obvious. They are hidden in the process.

According to Dario, a Panamanian contract clause that provides a completely seamless, bonded ocean-to-ocean/port-to-port transshipment process is the key to the system functioning as designed. They achieved this through the use of electronic data interchange (as once popularized, if you remember, by Ross Perot) to integrate the needs of government customs and agriculture requirements. This was truly revolutionary and allayed the concerns of the government, the railway, and customers alike.

Dario says, "Negotiating a contract clause that allowed us to build our own pier on the Atlantic, which in turn allowed us to economically and effectively bring in 283,000 tons of Nova Scotia ballast, 150,000 tons of concrete ties, 11,000 tons of new 136 lb. rail, all our [own] locomotives and rolling stock . . . made it work. The equipment was unloaded directly into our Atlantic concession area. The locomotives and rolling stock were unloaded directly onto dock side tracks connecting to the main line. This [innovation] allowed us to efficiently and completely rebuild the railroad to top world-class standards. The idea was to invest up front to minimize future maintenance and provide state of the art service with trains able to run with conveyor belt frequency because of the short length. Normally a railroad cannot compete with trucking for short haul distances. The concept of being able to haul huge volumes under a conveyor belt concept, coupled with seamless, bonded service, allowed the Panama Canal Railway to overcome conventional short haul obstacles."

The thrust stage was entered successfully and continues to provide returns for all involved to this day. In 2005 the Panamanian railway transshipped approximately 100,000 containers between the Atlantic and Pacific.

My hat is off to the entire team, companies, businesspeople, government officials, customers, and workers who make this engine hum through all the stages. This is Intelligent Innovation.

APPENDIX A: SRM AND ESRM RESEARCH QUESTIONS AND STUDY METHOD

In this appendix more detail is provided on the methods used to develop the Strategic Risk Management (SRM) Survey and the Extended Strategic Risk Management (ESRM) Survey.

STUDY LAUNCH

The study was commissioned after a particularly intense risk management training session performed for TTC Incorporated, where an audience member questioned the validity of risk management and then the validity of any management. His questions were designed to be provocative and controversial, not adversarial, but they were largely unanswerable with the current data available in the marketplace. As a result, a series of 15 questions were developed by Strategic Balancing Management Consulting's research division. The questions aimed to measure the decision-making efficiency (linked to requirements basis), innovative quotient (I-Quotient, or, as defined previously, an objective grade as to a company's ability to promote and capitalize on innovation), and use of vision and mission in management (a hidden measure of management focus, employee stress, and organizational health) of the respondents.

STUDY METHODS

In the brief example in Chapter 1, Jane was just one of hundreds of subjects and respondents and cases in the SRM studies. The mechanics of the study are described in the following.

In an effort to keep the purity of the study beyond question, a double-blind approach was used. Each questionnaire was kept totally anonymous and the results tabulated purely by demographics. No participant was told of the ultimate nature of the study; participants were told simply that it was a study on management and nowhere were they asked directly about innovation, decision management, or systems engineering usage. These topics were tabulated separately and linked to the results after the respondent filled out the study. For the primary study, the respondents were all employees in good standing with a minimum of one four-year BS/BA degree. Most had MBAs and MSs, and several had PhDs. All had been sent to an industry-wide conference at least once in the past year, with their organizations spending a minimum of $1,000 on their attendance. This is perhaps one of the most powerful aspects of the primary study. These are intelligent, productive people who have had exposure to other organizations and who are able (at least that was an assumption) to discern the nature of their organization relative to best practices and other institutions.

The study required over two years to complete. It was performed in person on three continents and included complete surveys from over 100 of the world's most revered firms from the European Community, Australia, the United States, Canada, and Mexico. Organizations included Mercedes-Benz, Ford, Boeing, NASA, General Electric, the U.S. Nuclear Regulatory Commission, Analog Devices, MIT, and others. Several much smaller firms were also included, and the respondents filled the same requirements as the larger firms. This data was used to form the I-Factor graphic, as well as provide information and trends on specific topics.

For the second survey, the ESRM Survey, the level of formality and the quantitative nature were toned down in favor of a more qualitative approach designed to ferret out broad trends or underlying procedural or other forces. The ESRM Survey used any person or organization that had experienced great failure or great success (preferably both) during a product development project or life situation. These study results were compared to the primary SRM Survey results, and some startling similarities surfaced. One basic hypothesis was that smaller or more entrepreneurial firms would have higher I-Quotients, higher risk tolerance, and greater success. Another basic hypothesis was that "adequately" funded (including machinery and people) projects would have higher success rates than underfunded or undersupplied projects. The conclusions were clear: Both hypotheses were wrong. Risk tolerance and an understanding of product development phases and subsequent in phase activities, in phase metrics, and in phase management techniques was by far the greatest contributor to success, regardless of size or "entrepreneurial" bent or funding. Risk tolerance, or at least an understanding of risk by phase, was important, but success was not necessarily linked to "high risk" or entrepreneurs or funding.

An excerpt of some of the questions from the SRM Survey follows:

Questionnaire for Research Project

Size of Your Organization:

❑ Small (0 to 100) ❑ Medium (100 to 1,000) ❑ Large (1,000 to 5000) ❑ Very large >5,000

Type of Organization:

❑ Manufacturing ❑ Service ❑ Public service (government, education)

❑ Medical (hospital, etc.) ❑ Defense

Your Position:

❑ Mid-level management ❑ Regular worker ❑ Senior management

❑ Decision Management

- How many times is each decision made in your organization on average?
  ❑ 0–2 (First decision plus one or two approvals)
  ❑ 1–3 (First decision plus two or three approvals, reworked decisions)
  ❑ 1–4 (First decision plus two to four approvals, reworked decisions)
- Do decisions have a clear requirements basis in your organization?
  ❑ Often ❑ Sometimes ❑ Seldom
- Do internal R&D projects submit an expected return on investment or some other proof or return?
  ❑ Often ❑ Sometimes ❑ Seldom
- Are development programs required to reapply for funding or return unused funding at the end of the yearly financial period?
  ❑ Often ❑ Sometimes ❑ Seldom

Innovation Quotient:

- When is the most "innovation" expected during a new product development program?

❑ Early concept phases ❑ Middle, detailed design phase ❑ Late construction phase

And so on.

The ESRM study included many of the original questions, plus an in-depth interview. These questions can be part of a holistic strategic planning process based on the vision for the organization. The vision can be decomposed all the way down through strategy to tactics and operations, and adjusted slightly to account for differences in target markets, profitability, and so on. This approach creates harmony and synergy between the various divisions of the company while retaining the unique competitive stance that has sustained you to date.

There are many areas where the methodology can help guide an organization during periods of growth while increasing the consistency and efficiency of decisions throughout the organization. On a more specific level, I was fascinated by the stage many of the respondents were in: emerging from the concept exploration and early product development phase to the preliminary design and solidification phase. This is an area where many questions come up and answering those questions with a strategic bent can really position the organization for future success. For example, attempting to answer the following three questions often spurred contentious debate among executives who just before the questioning were seemingly in alignment—on the same team and ready to move forward. These simple questions threw many teams for a loop.

- How do you develop brand recognition while leaving some room for products to solidify and morph in the coming year?
- How do you price products and services so as to develop a strong base while not selling yourselves short for the future?
- What is your target market? Is the market segmented, and which segment gets first priority while resources are constrained?

In addition, the survey included the following questions:

- Is your organization synchronized? Do disparate departments know what each other is responsible for and how each contributes to the common goal?
- Is your organization oriented in a single, explainable direction?
- Do the general manager and the CEO have roughly the same view of the goals of the firm?
- By what method will you establish corporate identity in the marketplace?
- How much of your total marketing budget will you use on Web page development? Why?
- How will you decide what your infrastructure investment will be in relation to other budgets?
- What method will you use this year to gain ground on your competitors, or if you are a nonprofit, how will you gain more funding?
- What percentage of your time is used in strategic decision making? What percentage of your time is used dealing with personnel issues?
- How is authority delegated?
- How is responsibility delegated?

- What are the one-, three-, and five-year goals of your firm? Do your employees know those goals? Do they know how those goals affect their daily job?
- By what method do you perform competitive analysis? Are you looking for a specific piece of information in the competitive analysis or for general trends?
- Is your firm successful consistently, sometimes, or never a one-hit wonder?
- Describe the best successes.
- What were the biggest factor(s) in these successes?
- Are you an engineer, software programmer, budget analyst, project manager, product manager, director, vice president?
- Do you currently manage multiple projects? Are the projects in various states of completion?
- Do you currently budget for R&D or production? Make decisions on asset allocation or capacity planning?
- How?
- Have you ever had to cancel a project because of lack of resources or funding or performance?
- How many meetings do your managers attend in a typical week? Do you think that is too many, not enough, or just fine? Do you, in general, think those meetings are productive? Could they be more productive?
- If someone were to attempt to acquire you, would you know the value of your firm? Would you have to think about whether or not to sell?
- Who are your competitors?
- Are you branding or gaining market share or both?
- How do you parse out retained earnings to various equally valid R&D projects?
- What is your succession and delegation strategy?
- How do you price products and services so as to develop a strong base while not selling yourselves short for the future?
- What is your target market? Is the market segmented, and which segment gets first priority while resources are constrained?

And so on.

APPENDIX B: CASE STUDY: SOLVING THE RISK VS. INTELLIGENT DILEMMA USING THE 3D RISK METHOD, SRM, AND ESRM RESEARCH

In this appendix, a case study involving the use of the entire 3D risk method is presented. It involves understanding traditional risk–return thinking, inverse risk–return thinking (also called opportunity management), and understanding the maturity phase, engine phase, decision management, strategy, marketing, and target market segmentation.

THE INTELLIGENT INNOVATION 3D RISK MANAGEMENT EXAMPLE

A particular program, the development of a new car design, requires that a revised transmission be mated to an existing power plant and to a totally new body. The three projects, involving transmission, power plant, and body, would have separate risk management programs (not separate risk management processes), which are summarized in a risk plan and applied by the appropriate work breakdown structure (WBS) element. We will also assume the three projects have different earned value management (EVM) and net present value (NPV) standards, which summarize to a program plan and are also linked to WBS elements.

The program can tolerate a certain level of cumulative risk, say a 6, which fits into the corporate portfolio. The portfolio consists of financial returns, liability concerns, warranty projections, profitability, and plant investment on 12 models. Each model is a program. Of the 12 models, which are always in flux, 2 are considered totally new, 6 are in various stages of revision, and 4 are established cash cows. Our new car program consists of various projects. The engine is one major project. The engine on our example car was specifically chosen for its proven reputation and known cost. For the first two model years, this old engine will be used while the actual engine for the new car is tested and produced in low quantities. The goal or strategy here is to ensure the model's overall acceptance and a known level of warranty costs and profitability. Even though the new engine has 26 fewer major moving parts, will cost less to manufacture, and will eventually be more reliable, it is still in development and considered unfit for first-run production with the stated goals.

The transmission will be revised with a new electronic control unit based on fuzzy logic and will have a breakthrough torque converter, while the remainder of the unit remains unchanged. These revisions to the transmission are considered risky; however, failure of either will result in performance penalties barely noticeable by most consumers. The strategy here is to test these innovations (at a high risk tolerance) on the economy car market and bring them to the luxury car market once proven, in effect reducing risk to the flagship luxury line. Reputation and warranty costs on the transmission are projected to be minimal if the two improvements are not fully reliable by introduction. The body structure and style is totally new. Several uncharted manufacturing methods are being developed to produce the complex body shapes and join them with fewer welds and fasteners than the previous model, in accordance with the corporate vision to become more leading-edge. With fit and finish high on any consumer list of satisfaction criteria, this is a high-risk venture. Some programs call the most critical goals "fence posts." Fit and finish would equate to a program fence post such as weight, cost, or delivery schedule. Delivery schedule is often a critical fence post because of marketing schedules. For example, the product delivery may need to coincide with the Super Bowl, where expensive advertising was scheduled to debut.

3DRM suggests that these are three totally separate risk management projects, requiring different consequences linked to their different strategies, which fold into the corporate strategic plan. Most of today's risk management efforts would apply stiff penalties (consequences, ROI, or NPV) to the body project, because of the cost of molds and line equipment, and low penalties to the engine program, because of the low value of already amortized equipment. Most of today's innovation practitioners would focus on new features for the car and leave out any look at assembly methods, tooling, and the like. These practices follow widely accepted accounting practices that also may run counter to the innovative needs of a development project and company. In this example, the engine is positioned to be the program's financial, risk, and reputation (quality) foundation. However, the engine would end up with the fewest "catastrophic" or "high" risk grades and mitigation plans because of the lower cost levels, small budgets, and minor component modifications. In a counterintuitive twist, 3DRM suggests that the engine, not the new body, should end up with the most grades in the red because of its strategic intent and phase position in the overall program. Traditional risk management would actually force results contrary to corporate goals if used on the engine.

Using 3DRM on the body yields the fewest red marks and maintains mostly green tracking marks through the detail design phase. The green grades of 3DRM replace items traditionally rated red, such as new technologies and processes. Since the organization as a whole has a goal to develop new processes, they are "encouraged" by the risk program, hence "Intelligent Innovation," a planned, deliberate fostering of innovative solutions.

Here we begin to see all four stages of the turbine model in action. Each stage is critical for the next, and the hand-offs, linkages, information, and strategic flow are what make it succeed or fail.

Another example of Intelligent Innovation and 3DRM is found in the engine mounting units. While the engine is largely unchanged, it did require a new mounting system for the new chassis. This mounting system should be marked as "catastrophic" for even a marginal likelihood of failure. In a traditional view, its budget is barely noticeable, say $500,000 in development and execution costs. On a $3 billion program, it is not even worth filling out a risk report. However, if the engine mounts do not integrate well with the new frame, undesirable vibrations can be transmitted to the passengers through the new and more rigid body structure. In effect, the least becomes the greatest in the detail design and construction phase. Since the engine is in the (final) construction phase, any risk carries a severe penalty with 3DRM, which normally places an emphasis on execution here.

Under traditional RM methods the new body structure is easy to rate high or catastrophic. Each fender mold, now used for thermoplastics, can cost $50 million. Each subframe stamping station costs $200 million. According to the typical risk management rating chart, almost any activity in the uncharted realm of the body project would be rated "high" or "catastrophic" even with a moderate likelihood of failure score. The high ratings may encourage the engineers to

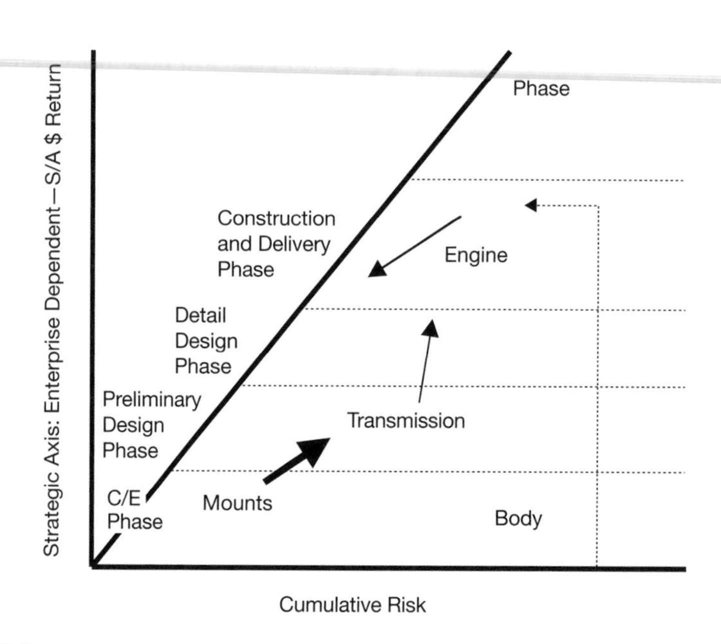

Figure B-1 Enterprise portfolio view, 3DRM example.

use existing tooling and methods to avoid short-term risk and/or cost and schedule penalties. This avoidance would sacrifice potential long-term production gains if new tooling and methods had been developed. Intelligent Innovation and 3DRM say that a problem with the fit and finish of the new body is considered acceptable for the first run because the overall benefit to the firm's other 11 models can be monumental. At the appropriate time or phase, the 3DRM risk curve rotates and applies gradually increasing consequences to a loose or unproven design solution. The modified risk management rating chart and the 3D portfolio graph, shown in Figure B-1, encourage intelligent, strategic risk (not unnecessary risk), possibly yielding significant long-term payoffs

For example, the body is located in the concept exploration (C/E) phase, but the arrow representing it is high on the y-axis because of the high potential return if it succeeds and it is far to the right on the x-axis (composite risk: Cp × Lp), reflecting its high overall risk. By contrast, the engine is in the construction phase, with a relatively low risk level and a return level that is medium and trending downward. When the projected risk-return quotient of the old engine equals that of the new engine, it's time to switch to the new engine.

APPENDIX C: PLAN OF ACTION AND MILESTONES

D		Task Name	Duration	Start	Finish	Predecessors
1		APPENDIX C : SAMPLE POAM	0 days	Fri 10/1/04	Fri 10/1/04	
2		**SAT SOFT V0**	**120.4 days**	**Fri 10/1/04**	**Fri 3/18/05**	
3		Draft SSDD Release 0 (R1.2)	33 days	Fri 10/1/04	Tue 11/16/04	
4		SSDD to Pubs	7 days	Wed 11/17/04	Thu 11/25/04	3
5		Send SSDD to Team, Final Review	6 days	Fri 11/26/04	Fri 12/3/04	4
6		Receive Comments, Finalize, Issue	7 days	Mon 12/6/04	Tue 12/14/04	5
7		**Componentsaka Modules**	**47 days**	**Fri 10/1/04**	**Mon 12/6/04**	
8		**SAT Msg Mdlw**	**47 days**	**Fri 10/1/04**	**Mon 12/6/04**	
9		**SAT OP ENV- CommonOperating Environment**	**47 days**	**Fri 10/1/04**	**Mon 12/6/04**	
10		Contractor 7 to fix basic functions	30 days	Fri 10/1/04	Thu 11/11/04	
11		Troubleshoot, Integrate, Compile	13 days	Fri 11/12/04	Tue 11/30/04	10
12		Document, Check SSDD	4 days	Wed 12/1/04	Mon 12/6/04	11
13		Work with Contractor 7 on Improvements	2 days	Fri 10/1/04	Mon 10/4/04	
14		**Service Agent Management(Sat Infra 1.1)**	**95 days**	**Fri 10/1/04**	**Thu 2/10/05**	
15		**Sat Infra- Real Time Agent Infrastructure**	**95 days**	**Fri 10/1/04**	**Thu 2/10/05**	
16		Debug for Version 0	30 days	Fri 10/1/04	Thu 11/11/04	
17		Begin Integration for V1	1 day	Fri 11/12/04	Fri 11/12/04	16
18		Troubleshoot, Integrate, Compile	10 days	Mon 1/24/05	Fri 2/4/05	13
19		Document, Check SSDD	4 days	Mon 2/7/05	Thu 2/10/05	18
20		**Simulation Support**	**47 days**	**Fri 10/1/04**	**Mon 12/6/04**	
21		Tele Workflow Modeling - Stand Alone	15 days	Fri 10/1/04	Thu 10/21/04	
22		TSB Gateway Support	14 days	Fri 10/22/04	Wed 11/10/04	21
23		aka (TSB link to Control)	1 day	Thu 11/11/04	Thu 11/11/04	22
24		Troubleshoot, Integrate, Compile	13 days	Fri 11/12/04	Tue 11/30/04	23
25		Document, Check SSDD	4 days	Wed 12/1/04	Mon 12/6/04	24
26		**Licence Issues**	**101 days**	**Fri 10/1/04**	**Fri 2/18/05**	
27		Research	10 days	Fri 10/1/04	Thu 10/14/04	
28		Solve	10 days	Mon 2/7/05	Fri 2/18/05	27

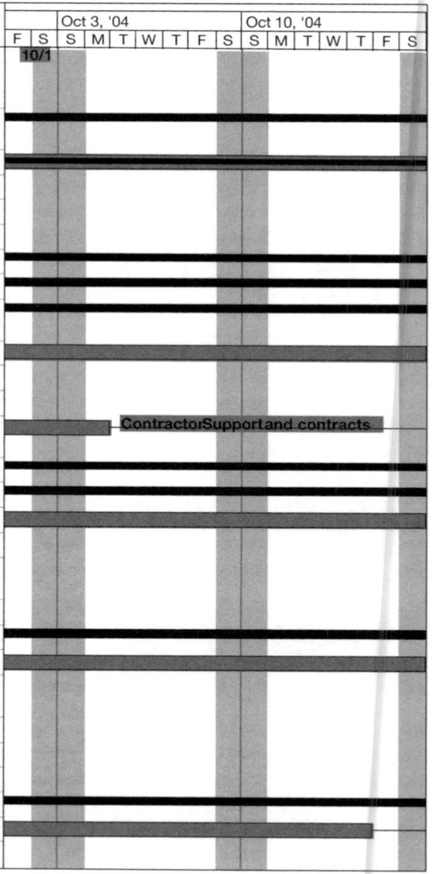

Oct 3, '04 — Oct 10, '04
F S S M T W T F S S M T W T F S
10/1
ContractorSupport and contracts

29		**Version 1 Start Up and Development**	**78.4 days**	**Tue 11/30/04**	**Fri 3/18/05**	
30		Planning Session with Team	1 day	Wed 12/1/04	Wed 12/1/04	
31		Revise Plan	4 days	Mon 12/13/04	Thu 12/16/04	30
32		Finalize Plan	10 days	Mon 1/10/05	Fri 1/21/05	
33		**Components,aka Modules**	**78.4 days**	**Tue 11/30/04**	**Fri 3/18/05**	
34		**Sat ControlCenter (GUI)**	**78.4 days**	**Tue 11/30/04**	**Fri 3/18/05**	
35		Develop GUI for Sat Console, Supports #44	35 days	Tue 11/30/04	Mon 2/28/05	
36		Troubleshoot, Integrate, compile	10 days	Mon 2/28/05	Mon 3/14/05	35
37		Document, Check SSDD	4 days	Mon 3/14/05	Fri 3/18/05	36
38		**Link Up Agent Management**	**70 days**	**Tue 11/30/04**	**Mon 3/7/05**	**13**
39		Directory Service Mgmt Sys View	14 days	Tue 11/30/04	Fri 12/17/04	
40		Develop (Sat Infra 1.1 W/directory services) Upgrade	21 days	Mon 2/7/05	Mon 3/7/05	18

Figure C-1 Sample POAM, page 1.

D		Task Name	Duration	Start	Finish	Predecessors	Resource Names	November B	B
41		Troubleshoot, integrate, Compile	10 days	Mon 1/17/05	Fri 1/28/05				
42		Document, Check SSDD	4 days	Mon 1/10/05	Thu 1/13/05				
43		**System PerformanceAnalysis and Optimization**	**48 days**	**Tue 11/30/04**	**Thu 2/3/05**		**Team**		
44		Performance Monitors - Agent and System wide Monitor (sub tasks) Bandwidth and Timing	1 day	Tue 11/30/04	Tue 11/30/04				
45		Develop Performance Monitoring (Bandwidth, Timing between Events	3 days	Wed 12/1/04	Fri 12/3/04	44	Moe and Curly,Larry		Moe and
46		Develop Sat Control Console (Geographic- terravision)	10.5 days	Mon 12/6/04	Wed 12/22/04	45	Contractor 4 and Interface Team		
47		Develop Sat Control Console (System Performance)	6 days	Wed 12/22/04	Thu 12/30/04	46			
48		Begin - Develop Sat Control Console (State/transition views) for V2	10 days	Thu 12/30/04	Thu 1/13/05	47			
49		Troubleshoot, Integrate, Compile	11 days	Thu 1/13/05	Fri 1/28/05	48	Team		
50		Document Check, SSDD	3 days	Tue 2/1/05	Thu 2/3/05		Team		
51		**SimulationIntegrationSupport**	**49 days**	**Tue 11/30/04**	**Fri 2/4/05**				
52		**Sat Integration(Control and Comm Integration)**	**49 days**	**Tue 11/30/04**	**Fri 2/4/05**		**Sammy**		
53		Simulation- Test and refine	15 days	Tue 11/30/04	Mon 12/20/04		New Person		
54		Look at interaction with other modules - Integration	20 days	Tue 12/21/04	Mon 1/17/05	53			
55		Integrate, Compile	10 days	Mon 1/24/05	Fri 2/4/05	54			
56		Doucment, Check SSDD	4 days	Tue 2/1/05	Fri 2/4/05				
57		**Controlintegration(done in V0)**	**25 days**	**Mon 1/3/05**	**Fri 2/4/05**		**Interface Team and Contractor**		
58		Begin Develop new Models, Update TSB Gateway and Integrate'	25 days	Mon 1/3/05	Fri 2/4/05				
59		Check current description of V0 Control integration in SSDD	15 days	Mon 1/3/05	Fri 1/21/05		Joe or replacement		
60		**Human IntegrationSupport, Gamepad Development**	**49 days**	**Tue 11/30/04**	**Fri 2/4/05**		**Sammy and Garage Folks**		
61		Game Controller	10 days	Tue 11/30/04	Mon 12/13/04				
62		Controller API	10 days	Tue 11/30/04	Mon 12/13/04				
63		World Map Integration, Compile, Troubleshoot	4 days	Tue 12/14/04	Fri 12/17/04	62			
64		Document, Check SSDD	3 days	Mon 12/20/04	Wed 12/22/04	63			

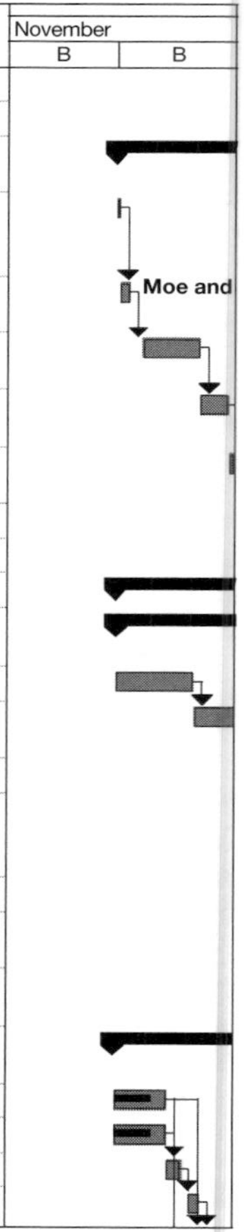

ID	Task Name	Duration	Start	Finish	Predecessors	Resource Names
65	Mouse Integration - Ongoing V1 and V2	27 days	Thu 12/23/04	Fri 1/28/05	61,64	
66	Begin 3D Performance Metrics (Old 65)	27 days	Tue 12/14/04	Wed 1/19/05	61,62	Contractor 4 and overflow team
67	Integrate, Compile, Troubleshoot	5 days	Mon 1/31/05	Fri 2/4/05	65,66	
68	Document, Check SSDD	4 days	Mon 1/31/05	Thu 2/3/05	65	
69	**Generic Message Template Definition**	**48 days**	**Tue 11/30/04**	**Thu 2/3/05**		**Core Team**
70	Field Types	11 days	Tue 11/30/04	Tue 12/14/04		
71	Representations	11 days	Wed 12/15/04	Wed 12/29/04	70	
72	Meta Data Model Wrap Up	10 days	Thu 12/30/04	Wed 1/12/05	71	
73	Troubleshooting, Integration, Test	9 days	Thu 1/13/05	Tue 1/25/05	72	
74	Documentation (ICD or IDD) so it can be used by Sat Areas	7 days	Wed 1/26/05	Thu 2/3/05	73	
75	Check SSDD	4 days	Tue 11/30/04	Fri 12/3/04		
76	**Memory Definition**	**51 days**	**Tue 11/30/04**	**Tue 2/8/05**		**ContractorX Solns**
77	Preliminary Memory Definition	7 days	Tue 11/30/04	Wed 12/8/04		Contractor X,Sally Jo
78	Initial bSat Infran storming meeting with Contractor 4 and MI Team	1 day	Thu 12/2/04	Thu 12/2/04		Team
79	Document what it will do, how constructed, tagging modules	4 days	Fri 12/3/04	Fri 12/10/04	78	Contractor X,Sally Jo
80	Finalize V1, V2 Memory definition	36 days	Fri 12/10/04	Mon 1/31/05	79	
81	Insert description into SSDD	9 days	Thu 1/27/05	Tue 2/8/05	80	Jimmy Dean and Donna,Jimmy Dean
82	**Ground Cntrl Lab Set-Up**	**38 days**	**Tue 11/30/04**	**Thu 1/20/05**		**Team**
83	Overall Coordination	3 days	Tue 11/30/04	Thu 12/2/04		Carl,John
84	Get all Software together	5 days	Fri 12/3/04	Tue 12/14/04	83	Moe,Larry
85	Get all machines going, attached	5 days	Fri 12/3/04	Thu 12/9/04	84	Sammy,Team,Candy
86	Install	8 days	Thu 12/9/04	Mon 12/20/04	85	Moe,Larry
87	Check on security, foreign code, freeware licences	30 days	Fri 12/10/04	Thu 1/20/05	85	Proj. Mgmt
88	Troubleshoot	6 days	Tue 11/30/04	Tue 12/7/04	85	Moe,Larry
89	Finalize	1 day	Tue 12/21/04	Tue 12/21/04	86	Moe,Larry
90	WContractor 4 procedure and Lessons Learned for MI Knowledge Base	1 day	Mon 12/20/04	Mon 12/20/04	86	Moe
91	Announce, Schedule Demo	1 day	Tue 11/30/04	Wed 12/1/04		Proj. Mgmt

Figure C-2 Sample POAM, page 2.

APPENDIX D: WORK CONTENT

Additional information on the elements of work content and capacity planning for a more thorough analysis and POAM development is provided in this appendix.

DEFINITIONS

Work content is the volume of work represented by a particular function/part/activity in its unfinished form. An alternative definition is that work content is the actual amount of time a particular function or job will consume of a particular resource for a defined operation.

Work center capacity is the amount of product that can be produced by a given work center with a given efficiency, uptime, break time, and so forth. Planning, at all levels of detail, requires knowledge of a work center's capacity and schedule as well as the capacity and schedule (and cost of operation) of alternate work centers to make complete decisions. Process owners should be in agreement with the data and procedures used by these planners regarding each work center.

Work content includes the following:

- Setup time (amortized if multiple parts)
- Reasonable intra work center transport

Work content does not include the following:

- Personal break time

- Other legal/accounting/administrative overhead unless directly attributed to the activity
- Transport between major operations
- Machine downtime or utilization
- Labor considerations

Work content is separate from work center capacity, which in turn is separate from sequencing. They must be used in conjunction (with cost information as well) to provide a meaningful data set to planners, forepersons, and make/buy decision makers.

TIME: THE ONLY COMMON METRIC

Work content calculations can change depending on the work center in question. Many companies use average work content or best-practice work content for scheduling and costing considerations and avoid actual individual calculations. Work content is not weld linear feet, number of pipe bends, number of nc programs, and so forth. Instead, it is the time those operations require. Time is the only measure that transcends various operational needs (planning, scheduling, sequencing, accounting, costing, etc.).

For the capacity planning and analysis functions, your company needs to work in an integrated manner. There must be a common set of assumptions and conventions used by engineering, accounting, planning, and budgeting functions. If common assumptions and conventions are not used, there will be a conflict when supply and demand curves are aggregated because they are not based on the same units (demand) and cannot be allocated to like work centers (supply).

Capacity is generic. For example, for a capacity planning system to work correctly, a CAD designer is a CAD designer (with some obvious constraints) and has a certain capacity. Someone who understands weights calculations is just that. It does not matter (in theory) in which group, division, or location that capacity resides—especially in today's global economy.

SCHEDULING IS A KEY ENABLER

The scheduling of the resource does matter to its manager. He or she is responsible and must balance various demands against the total capacity by moving jobs in space or time or to an alternate capacity pool. However, to the people creating the demand, the task requester, capacity, location, overtime premiums, and the like really do not matter as long as the "capacity" is qualified to do the job. The demand simply concerns obtaining a certain product, at a certain quality level, at a certain cost (work content in hours), and by a certain point in time to support a network activity that supports the overall build strategy. Since all activities are linked in the network and the network decomposition is what creates the demand (typically through the manufacturing resource planning system), a department (resource ownership) is rather inconsequential. What is consequential is resource balancing and on-time delivery of products and services. Again, this is where a common definition and coordinated work center identification comes in—one that is linked to a POAM and linked to a vision and a mission. The division level is greater than the department level—that is, the needs of the many outweigh the needs of the few, which is an argument for centralized planning. The central planning case is based ultimately on strategic focus, individual task margins, and the concept of balancing the profit margins against the long- and short-term cash needs of the company.

In more advanced cases an algorithm is developed that provides the capacity balance with the financial ramifications of each move that is made. This financial information is in addition to the information on ramifications of capacity moves on build strategy, customer satisfaction, strategic weights (for example, if project A may lead to a large highly visible future contract with NASA, it may take precedence over project B for a small unknown firm), financial profitability weights (for example, if project A has a big early completion bonus, it may take precedence over project b), and so on. All of this thinking only works if there is a tight, complete corporate strategic plan linked to a common vision and mission statement.

APPENDIX E: I-QUOTIENT DETAILED METHOD

THE EXPANDED SELF-EVALUATION

In this section an outline is supplied for those who wish to perform a more advanced I-Quotient self-evaluation. The outline lists some of the many business functions that can contribute to or detract from achieving high performance.

There are two ways to use this outline. The first is to simply use it as a checklist for areas to investigate further after using the self-evaluation tool. The outline is used by first taking the basic self-evaluation test and discussing the potential areas that are the root cause of the real or perceived gaps within the organization. For example, suppose an organization received mediocre scores and there is a consensus among managers that the cause of less-than-world-class innovation is poor employee retention. The company strives to be innovative and has made significant measurable attempts to remove process and policy impediments, but employees leave saying that they enjoyed working there but left because they were given better offers elsewhere. This may lead to a project centered around category 1, "Resource supply," and subcategory 1.1.4, "Benefits." Benefits for this organization could be compared to the average in the industry and adjusted if necessary to improve employee retention. If the benefits were found to be satisfactory, then another suspect area could be studied, and so on.

The second way to use the outline is as an in-depth evaluation tool. Take each category and subcategory and give it an I-Quotient score using the preceding criteria. Our database does not have this level of delineation in the data; however, some generalized aggregate scores are presented for discussion purposes. Perhaps more important are the discussion and reflection necessary to score each category and subcategory. The value in that type of introspection will help any organization pinpoint areas of conflicting opinion—often a harbinger of a business function that is not well understood within the organization. This exercise will also often pinpoint areas of vigorous agreement among the graders. If everyone agrees for the better, assign a good grade in a specific area, then celebrate the good news and move on to areas that need more help. If folks agree that a particular area has significant weaknesses, then set to fixing it.

If a user has nothing pertinent in a subcategory, such as 1.1.2, just aggregate up to the score of the next highest category. If an entire category is not applicable, such as "consumables," use a middle score such as a 5, which will provide an adequate level of accuracy. Other users often use an average of all their applicable scores as a placeholder for areas that were not applicable.

SELF-EVALUATION OUTLINE

Major I-Quotient detailed self-evaluation areas are as follows. The test and analysis can be applied in one or all of these areas as needs dictate.

1. Resource supply
    - 1.1. Labor pool
        - 1.1.1. Hiring, recruitment strategies
        - 1.1.2. Firing
        - 1.1.3. Training
        - 1.1.4. Benefits, including educational reimbursement and other fringe benefits
        - 1.1.5. Motivation, including fast tracking, cross training, etc.
        - 1.1.6. Advancement
        - 1.1.7. Force balancing, long-term forecasting, etc.
        - 1.1.8. Recruitment
    - 1.2. Product pool
        - 1.2.1. Plant
        - 1.2.2. Property
        - 1.2.3. Equipment
        - 1.2.4. Maintenance methods, funding, strategies
        - 1.2.5. Planning and line balancing
        - 1.2.6. Finite capacity planning
        - 1.2.7. Component acquisition, methods, and strategies

- 1.2.8. Outsourcing
- 1.2.9. Just-in-time production
- 1.2.10. Lean manufacturing
- 1.2.11. Etc.

2. Resource demand
   - 2.1. Sales
   - 2.2. Marketing
     - 2.2.1. Traditional print and media
     - 2.2.2. Internet, including blogging, affiliates, links, comarketing
   - 2.3. Forecasting
   - 2.4. Strategic planning
   - 2.5. Partnering
   - 2.6. Acquisitions
3. Operations management
   - 3.1. Financial accounting
   - 3.2. Financial planning
     - 3.2.1. Debt management and strategies
     - 3.2.2. Equity management and strategies
     - 3.2.3. Tax strategies
     - 3.2.4. Retained earnings and stock price management
     - 3.2.5. Capital
   - 3.3. Asset management and planning
     - 3.3.1. Plant, property, and equipment
       - 3.3.1.1. Software licenses
       - 3.3.1.2. Other licenses
     - 3.3.2. Lease vs. buy
     - 3.3.3. Benefit funding
     - 3.3.4. Outsourcing and key tooling lease
     - 3.3.5. Intellectual property
       - 3.3.5.1. Ownership
       - 3.3.5.2. License
       - 3.3.5.3. Strategies
   - 3.4. Consumables
     - 3.4.1. Utilities
       - 3.4.1.1. Electricity and gas
       - 3.4.1.2. Water and sewage
       - 3.4.1.3. Fleet maintenance
       - 3.4.1.4. Etc.
     - 3.4.2. Materials
       - 3.4.2.1. Long lead time
       - 3.4.2.2. Commodity

4. Planning
   4.1. Vision
   4.2. Mission
   4.3. Long-term strategic planning
   4.4. Tactics
   4.5. Long-term operational planning
   4.6. Competitive analysis and planning
   4.7. Eternal market forces—analysis and planning
   4.8. Acquisition analysis
   4.9. Force gap analysis
   4.10. Production capacity gap analysis
   4.11. Key technology gap analysis
   4.12. Partnering
   4.13. Government regulatory planning
   4.14. Global economics
   4.15. Disaster planning
   4.16. Insurance management
   4.17. Pension funding and planning
5. Design and development
   5.1. Design tools
   5.2. Design methods
   5.3. Training and certifications
   5.4. Goals and metrics, quality
   5.5. Industry partnerships, work share, IP sharing
   5.6. R&D management
   5.7. R&D planning
   5.8. Laboratory and prototyping
   5.9. Design review and approval

TIPS ON USING THE I-QUOTIENT TEST

Step 1. Decide How to Take the Test

Middle managers and mid- to senior-level technical people provide the best results here if your organization can only afford the time to do it once—the theory being that the newer employees just don't have enough perspective to provide a relative measure and the senior people have too much perspective and are stovepiped. These are vast and unfair stereotypes, and if the organization has the time and resources, we suggest three independent, simultaneous tests for three groups: (1) new, inexperienced employees, (2) middle-level managers, and (3) long-time senior managers. The results are compared and clues as to perceptions of what is working and what is holding the firm back are discovered. I use the word "perceptions" because this is a qualitative analysis tool. It has a significant amount of science, data, research, and math behind its "engine" and results. However, the actual grading is by people based on their opinions and best guesses at that moment. So while it blends both quantitative and qualitative, it must ultimately be seen as a qualitative tool, capable of providing great insight into organizational trends, strengths, and weaknesses and capable of provoking healthy discussions but not capable of providing investment analysis, hiring–firing analysis, go/no-go analysis, and so on. More traditional marketing, finance, legal, and technical tool sets must be used for such decisions.

Step 2. Decide Where to Apply the Test

Different industries drive innovation in or out from different areas. Many drive innovation from the resource supply side, many from the operations management side, and some from all over. If possible, we recommend a multiaspect approach. The bottom line is this: If you are a customer- or consumer-driven industry, try the operations management and marketing sides first. If you are in an industrial-supply-, engineering-, or technical-driven market, try the resource supply side first.

THE ONE-YEAR PLAN FOR THE I-QUOTIENT SELF-EVALUATOR

Following is a rough idea of how to use the Self-Evaluator in an organization over the period of a year. Each organization should determine what is the best method for its unique purposes. The plan shown is designed for a midsized for-profit firm.

Quarter 1: Global Evaluation

Test a cross section of the entire organization to provide a global evaluation, and draw conclusions as to policies and processes that are affecting innovation, pro and con. Then test individual divisions or departments and tabulate those results. Compare and contrast the various scores against the global evaluation, looking for the sensitive area—a type of macro sensitivity analysis.

Next, develop an initial plan to improve innovation based on the I-Quotient results. Perhaps you need more talented engineers, or perhaps your pension policies are so deficient that your middle managers are jumping ship and the project management aspects of your firm have grown weak. Perhaps your financial accounting policies are so strict that no R&D project ever gets funded past the first year because of "failure."

There are many combinations and permutations of possible reasons that innovation is or is not flourishing in any organization. I can only provide a few brief simplified examples here regarding the type of analysis one can do on the data. The reader is encouraged to visit the Intelligent Innovation Web site (www.intelligentinnovation.com) and look for various downloads, workbooks, and tools expected to be developed to assist in the evaluation and recommendation aspect of using the I-Quotient Self-Evaluator. Similar resources and both free and fee-based tools can be found at the publisher's WAV site at www.jrosspub.com.

Quarters 2 to 4: Individual Evaluations

Individually analyze each major foundational area of the organization and juxtapose the results with the global test. Refine your improvement strategies with the deeper knowledge gained during the individual foundational area analysis.

Quarter 4: Metrics Evaluation, Self-Reflection, and Closure

In the fourth quarter and into the first quarter of the next year, it may be useful to take a step back and look at what worked and what did not work. This is a necessary step for those organizations that want to be world-class performers. Metrics are required that grade the various efforts as to each effort's original goal. The metrics must be accurate, relevant, and effort specific and provide telling information useful for decision making and resource management. For example, a typical metric may be employee retention before and after the change of a major benefit. That metric would be accurate enough, but to be relevant, it must also include the results of exit surveys of those people who left before and after the change. To be really good, that metric would then be compared to surveys of employees that are currently employed. These anonymous surveys would provide a backdrop or data to make the overall evaluation more powerful. Finally, the gaps, overlaps, and issues would be addressed one by one as prioritized by the organization's goals and vision and mission.

APPENDIX F: SUMMARY OF I-FACTORS

I-Factor 1: Any success, personal or business, is rooted in good decision making. Innovative decision making separates the wildly successful from the rest.

I-Factor 2: A key aspect of success is the willingness to look inside and outside your particular industry.

I-Factor 3: Innovation must be constant and everywhere in your process.

I-Factor 4: Decide what is worthy.

I-Factor 5: Decision-making efficiency, accuracy, and relevancy are all crucial to the economic and emotional well-being of the organization and are directly related to the dissemination of a well-crafted vision and mission.

I-Factor 6: Begin with the end in mind, communicate it to everyone, and set interim milestones that drive you there.

I-Factor 7: An attitude of resourcefulness can overcome aggressive milestones and resource/material shortages.

I-Factor 8: Turn limitations into opportunities and understand your organization's capacity so that an innovative approach to the opportunity can be capitalized on.

I-Factor 9: Understanding risk organizationally and strategically, by phase, can promote ingenuity and performance.

I-Factor 10: Good new product design, development, and fielding, in the long term, sustain corporations.

I-Factor 11: A culture of innovation cannot exist in a policy-driven atmosphere of managed risk and artificial accountability.

INDEX

G

H

I

J

K

L

M

N

S

- Identify
- Assess (Cf * Pf = Rf) (Consequence of failure * Probability failure = Risk factor)
- Mitigate
- Track
- Close

10. Summarize the total risk exposure using the 3D enterprise portfolio graph. Eliminate or combine duplicates, adjust conflicts, and mitigate unreasonable exposures.
11. Continue, refine, and adjust.

Many contemporary risk managers are finding that application of risk management to their organization requires constant monitoring. It requires encouragement of early team involvement, training, and approval and reapprovals by top management. In fact, effective risk management requires that the organizational processes are changed for true integration and success. 3DRM is no different and can be more difficult to apply if the proper groundwork is not applied. Since 3DRM is a function of the organization's strategic plan and goals, it should be inherently integrated with senior management's desires and other processes from the start, which will facilitate its acceptance. 3DRM, like regular risk management, should be promoted as a proactive, positive influence innovative management tool. It is a profit-producing activity, not a cost or chore.

CHAPTER REVIEW

Innovations can be high risk, low risk, and everything in between. They must be managed along with all the other activities and investments going on at the time. However, because of cultural and historical norms within a company, managing almost any innovation often gets bundled with managing high risks, regardless of the actual level of risk or the appropriate expectations for that phase of the development. 3DRM provides a method and, more broadly, a framework for assessing all risks with a phase-appropriate, strategically targeted consequence factor. This method includes the understanding of the need for failure on the way to success (analogous to the bypass in a turbine process) and allows for opportunity costs of not taking risks. Once the consequences for an organization are customized with regard to the vision, mission, and strategy of the organization by phase or maturity level, all risks and innovations can be parsed and managed for optimum performance.

MAKING IT REAL

Follow these steps and refine and reiterate as necessary until your organization has developed the correct set of parameters:

1. Train employees on the basic nature of risks and risk management. Warn them about the human nature tendency to rate all risks high, regardless of actual risk.
2. Have management take a risk tolerance test. Note your organizational risk distribution and conduct discussions to uncover the various risk profiles. Help the organization understand who is risk averse and who is risk seeking so that people will better understand one another's behavior and approach to various projects.
3. Develop specific consequence factors for your organization by phase (whatever phases are appropriate for your industry). Be sure to craft these consequence factors to include your strategic intents for each phase, keeping an eye toward propelling innovations through the Innovation Lifecycle. Try making the consequences slightly less catastrophic at each phase to encourage innovation throughput in non-critical (non-life-threatening) industries, and try doing the opposite in critical industries such as medical and transportation.
4. Commit enough funds and manpower to perform a quarterly or yearly risk portfolio. Rate all of the projects and programs and developments in your pipeline using the specially crafted consequence factors. Put all of this information in the 3DRM graph and analyze the portfolio. Look at your "sweet spot." Is it where you expected? Is it where you desired? Look at the areas where you have few projects under development. Are the number and level of risk what you expected? Are there enough projects at each phase to keep the engine running?

A customized portfolio management mentality and process should emerge. The organization's comfort level with the terms, tools, and theories will take some number of months and iterations to raise to an acceptable level. However, the adjustment is worth the trouble. This method can alert people to blind spots and judgment errors that typically occur when only looking at a project in two dimensions.

15

STOCKING UP ON INNOVATION

"By 2010, products representing more than 70 percent of today's sales will be obsolete due to changing customer demands and competitive offerings."

—***Deloitte Research LLC Global Benchmark Study*** *"Mastering Innovation,"* ***Exploiting Ideas for Profitable Growth****, March 2004*

So why bother? Why go through all this pain, thinking, and expense to bring new products and services to market? Why expend company resources, scarce R&D funds, and valuable production capacity for new products that are unproven and high risk? Why stress already tired employees to solve new problems? Why ask financial support people and bankers to extend yet another line of credit? Really, aren't these the ultimate questions? A few of us invent for the love of science or for the thrill and pride associated with forming and birthing a new idea. Call it the entrepreneurial spirit, but still, none of that thrill, pride, or spirit pays the bills, keeps the creditors at bay, makes the shareholders happy, or keeps the customers coming back. This chapter discusses the direct influence of innovation on a firm's valuation in both the financial market and the market of public opinion.

Answering "why bother?" brings us to our next I-Factor:

I-Factor 10: Good new product design, development, and fielding, in the long term, sustain corporations.

In this chapter, we* will provide proof of factor 10.

No amount of financial wrangling, merger mania, or technology hype can replace the benefit of consistent new product introductions based on innovation in both product and process. In other words, the engine cannot function without fuel and air.

This time I was leaving Amsterdam en route to St. Louis in the United States. Besides thinking about getting out of consulting because I was tired of airplanes, I was thinking of my mysterious friend and how she was doing. I had worked on my manuscript for months and was nearing a close. This time I had a question for her. I would deliver it deftly, like a Japanese master chef cutting the good meat from the jowls of the poisonous blowfish.

Plugging into the wireless airport free WAN, I see an e-mail from her consulting company. The e-mail read simply, "So?" She likes to be so mysterious and brief. I answered simply "Got it all. You need to buy the book to hear the rest! But one more question this time for you: Why bother?"

She answered back immediately: "Value." Value? One word! I ask her the meaning to life and she sends a single word! I sighed. This is getting annoying.

I embarked on a narrow study to answer this very question: Why bother? Why go through the risk, expenditure, heartache, and trouble of bring new innovative products and services to the world. What value does all this have? The stock price versus innovation study disclosed a connection between overall corporate performance and product development. The study is visible, verifiable, and repeatable. It covers over 15 years of public data in two industries. The results indicate that product design (which includes the actual design, the quality of the product, performance, price, etc.) alone is not sufficient for improved stock price (one measure of overall corporate performance), but that stock price growth does not occur unless significant (new and innovative) products are being consistently developed and fielded. There is a subtle and very important distinction in this conclusion. It tells us that even very good products reach a saturation point, particularly with the stock investment community. Their valuation is included in the valuation of the company already. So even a great cash cow, a high-performing,

* In this chapter, I am deeply indebted to Mr. Paul Sabin, a talented engineer, designer, and research assistant, who helped provide a substantial portion of the research that forms the basis of this chapter.